LIVING CONTROL SYSTEMS

About CSG

The Control Systems Group, Inc. (CSG) is a membership organization supporting the understanding of cybernetic control systems in organisms and their environments: *living control systems*. Academicians, clinicians, and other professionals in several disciplines, including biology, psychology, sociology, social work, economics, education, engineering, and philosophy, are members of CSG. Annual meetings of the Group have been held each fall since 1985. CSG publications include a newsletter and a series of books; a journal is being planned. For more information about CSG, write to The Control Systems Group, Inc., Business Office, 1138 Whitfield Rd., Northbrook, IL 60062.

The CSG logo, designed by Mary A. Powers and Gregory Williams, shows the generic structure of cybernetic control systems. A Comparator (C) computes the difference between a reference signal (represented by the arrow coming from above) and the output signal from Sensory (S) computation. The resulting difference signal is the input to the Gain generator (G). Disturbances (represented by the black box) alter the Gain generator output on the way to Sensory computation, where the negative-feedback loop is closed.

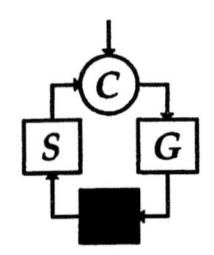

LIVING CONTROL SYSTEMS

Selected Papers of William T. Powers

Publication History

Second printing: January, 2005

Benchmark Publications Inc.
New Canaan, Connecticut 06840

ISBN 0-9647121-3-X

1 2 3 4 5 6 7 8 9

Printed in the United States of America

First printing: 1989

Production coordination: Gregory Williams, Gravel Switch, Kentucky
Printing: Diamond Graphics, Lexington, Kentucky

Typeset using *PowerForm* and *PictureThis* on a Zenith Z-100 computer, with camera-ready copy output produced by a NEC Silentwriter LC-890 printer.

Library of Congress Catalog Number: 89-62534
Original ISBN: 0-9624154-0-5

Copyright 1989 The Control Systems Group, Inc.
All Rights Reserved

1 2 3 4 5 6 7 8 9 10

Contents

Foreword, *Richard S. Marken* — vii

Preface, *William T. Powers* — xiii

A Note on the Text, *Gregory Williams* — xix

A General Feedback Theory of Human Behavior: Part I — 1

A General Feedback Theory of Human Behavior: Part II — 25

A Feedback Model for Behavior:
Application to a Rat Experiment — 47

Feedback: Beyond Behaviorism — 61

Behaviorism and Feedback Control — 79

Applied Epistemology — 89

The Cybernetic Revolution in Psychology — 103

Quantitative Analysis of Purposive Systems:
Some Spadework at the Foundations
of Scientific Psychology — 129

A Cybernetic Model for Research in Human Development — 167

Degrees of Freedom in Social Interactions — 221

vi Living Control Systems

On Purpose	237
Control Theory and Cybernetics	245
The Asymmetry of Control	251
An Outline of Control Theory	253
Published Works by William T. Powers on Living Control Systems, 1957-1988	295

Foreword

Some of the best science is done by people who refuse to take the obvious for granted. Copernicus didn't take the sun's daily trek across the sky for granted, Einstein didn't take the regular tick of time for granted, and William T. Powers didn't take the appearance of behavior for granted. The results of not taking things for granted can be powerful new ways of looking at the obvious, but the value of the new point of view is rarely appreciated immediately. Gregor Mendel, who didn't take the blending of hereditary traits for granted, approached the study of heredity in a completely new way, using a combination of botany and mathematics. For his creativity and hard work, having single-handedly invented the field of genetics, he was rewarded with complete and utter neglect. His work was finally recognized 30 years after his death (small comfort to Gregor) by three scientists who independently rediscovered his laws.

Mendel's story illustrates a rule of scientific discovery that is too often followed: "... look at a problem from a totally new angle and people won't so much disagree with you as completely misunderstand you. They won't grasp what you are talking about and will ignore you." (Maitland A. Edey and Donald C. Johanson, *Blueprints*, Little, Brown and Co., Boston, Toronto, and London, 1989, p. 105) Powers has looked at the phenomenon of behavior from a totally new angle and, sure enough, people have misunderstood him and ignored him, but they have rarely disagreed with him. The lack of disagreement is rather surprising, since Powers' ideas about behavior contradict the fundamental assumptions of scientific psychology.

Conventional psychology views behavior as evoked motor output; Powers argues that behavior is controlled perceptual input. These approaches to behavior could hardly be more different.

Powers developed his ideas about control after taking a close look at a phenomenon that most psychologists have taken for granted—behavior. The conventional wisdom in psychology is that behavior is what organisms do. This seems obvious. Behavior appears to flow out of organisms like water from a spigot or printout from a computer. It is difficult to believe that this view of behavior could be wrong, but it is. Powers saw that behavior is not produced solely by the organism, but by the organism in concert with unpredictable and usually undetectable environmental disturbances. These disturbances are pervasive but difficult to notice because behavior is ordinarily quite consistent. Organisms weave webs, migrate to specific destinations, build dams—and they do these things over and over again. Powers, looking at behavior through the eyes of a trained physicist and engineer, saw that such consistency was quite surprising. He realized that organisms can produce consistent results (a web, a landing in San Juan Capistrano, a dam) only by continually adjusting their actions to compensate for disturbances. It is as though they intend to produce these results and vary their actions appropriately in order to do so. Organisms seem to behave on purpose.

Psychologists before Powers had noticed the purposiveness of behavior. They saw, for example, that organisms produce consistent results using highly variable actions. But most psychologists ended up attributing this variability to "statistical noise"; Powers, on the other hand, saw it as essential. If actions did not vary, behavioral results would repeat only by chance, fluctuating as a result of the random effects of environmental disturbances. Instead, actions vary to compensate for the effects of disturbance, producing consistent results in an inconsistent world—a process called "control." Powers realized that the events we call behavior, from lever presses to religious ceremonies, are controlled results of action; to behave is to control. This was a momentous observation, and it needed an explanation. How were organisms able to control? Fortunately, the basic answer had already been discovered—it was control theory.

Powers is not the inventor of control theory. Nor is he the first to have applied it to behavior. He is, however, the first to have used control theory to explain behavior as an example of the phenomenon that control theory was designed to explain —control. Previous attempts to apply control theory to behavior put the cart before the horse, so to speak. People were more familiar with the theory than with the phenomenon of control. Thus, control theory was applied to behavior before anyone had any idea that behavior involved control. This was a bit like trying to develop a theory of evolution before there was any evidence that evolution had occurred. It was bound to lead to confusion and disappointment. Before Powers came along, control theory was already on the wane as a model for behavior. And for good reason. Control theory is the wrong model of behavior if behavior is evoked motor output. But it is the right model of behavior if behavior is control.

Powers built a model of behavior based on control theory. The basic tenet of the model is that organisms control perceptual input, not motor output. This is a fact of control system operation. Control systems act to keep their perceptions matching reference images of what those perceptions should be. They do this by acting on the environment, producing effects which, when combined with prevailing environmental disturbances, produce the desired perceptions. Living control systems are no different than any other control system in this respect. When we watch the behavior of organisms, we are watching living control systems "from the outside"—systems that are controlling their own perceptual experience. Behavior is, as Powers put it in the title of his classic book on the subject, "the control of perception."

To understand the behavior of a living control system, the observer must learn what perceptions the system is controlling: what reference images the system is trying to match. Living control systems produce many results, some of which may be controlled and others not. The observer must learn which results correspond to the perceptual variables that the system is actually controlling. These results are called controlled variables. Powers has described an objective method, called "the test for the controlled variable" (or simply "the test"), for finding out what variables a control system is controlling. When applied to living control systems, "the test" constitutes a new research methodology for psychology, with a new goal—the discovery of controlled variables.

Powers understood that the variables controlled by living control systems can be quite complex, as evidenced by the complex behaviors produced by organisms—behaviors like building skyscrapers and writing piano concertos. As a control engineer, Powers knew that a control system could be designed to do anything that it could perceive. Thus, a control system could produce complex behaviors if it could perceive complex variables. Powers showed how a hierarchy of systems controlling different classes of perceptual variables could produce the kind of complex behavioral results produced by living organisms. The model has a satisfying consistency with what we know of the functional and structural organization of the nervous system. Whether or not it proves to be completely correct (it has survived numerous tests, but there are a great many more to be done), it has served its purpose by showing that all behavior, from tensing muscles to writing poems, can be modeled by control theory.

A number of scientists, impressed by the power and beauty of control theory as applied to behavior, have devoted their research efforts to testing and expanding Powers' ideas on living control systems. Obviously, I am one of them. I knew after reading *Behavior: The Control of Perception* that Powers had something very important to say; I just wasn't sure what it was. It isn't easy to understand control theory at first reading, especially if you are thoroughly imbued with the concepts of conventional behavioral science, as I was. It takes a while to understand that control systems compensate for disturbances rather than respond to stimuli; that stimuli are controlled and not in control; that living control systems control and cannot be controlled.

In order to get a grasp of the control model, I sought out other works by Powers. I was able to find them, but it wasn't always easy. Now it is. You no longer have to be a fanatic to obtain Powers' finest publications: they are gathered together in this book. You can learn a great deal about living control systems by reading this book. But don't expect to find the answers to all your questions about life. Control theory provides a new foundation for the study of living systems, but it is just the foundation—it is not the edifice. A great deal of work must be done to build on this foundation, but construction can progress with confidence because the foundation is solid. Behavior *is* the control of perception. Understand that, and you understand the basic organizing principle of living systems. The rest, as Einstein said, are details.

Richard S. Marken
Los Angeles
July 1989

Preface

For uncomfortably close to 30 years I have been writing an article called "Control Theory for the Life Sciences." I have published this article in books and journals, newsletters and proceedings. It has been aimed at behaviorists, cyberneticians, linguists, biologists, social scientists, and anyone else I thought had a glimmer of interest in the subject. I have spoken this article to seminars of graduate students in several disciplines, to medical students, and to faculty members, in classrooms, lecture halls, and brown-bag lunchrooms. You will see the article here, in most of its incarnations. This is getting extremely tiresome, not only for those who have heard the message too many times, but for the one who has heard it the most often of all: me.

This persistence was not in the original plan, which was to communicate the basic theory worked out by R.K. Clark, R.L. McFarland, and myself in the 1950s and published in 1960, then to find a place to work and develop the basic ideas into a full-blown discipline. The theory was so elegant, so close to being self-evident, so clearly useful and explanatory, that neither I nor my collaborators anticipated any problems in gaining support for it. We knew that others were moving in the same general direction and thought they would welcome real signs of progress. I look back now and wonder how we could have predicted the future so poorly.

It is now clear that a new theory is quite welcome in the sciences of life, but *only if it does not call for revision of important beliefs*. It's all right, for example, to propose a theory saying that an organism's susceptibility to reinforcement by food might be modified by the use of water deprivation. It is not all right to propose a theory that says, in effect, that there is no such thing as reinforcement. There are simply too many scientists who rely on the concept of reinforcement as their main explanation of behavior, the foundation of their theorizing; take away that tool and they have nothing left. The same thing would happen to control theorists if the principles of control were shown to be spurious. Every science needs explanatory principles on which it can rely; if the principles of a science were reorganized yearly, no organized concepts could ever develop. I was naive to think that control theory could become influential in the life sciences over a period of a year or two, or even a decade or two. No idea that can change the course of a science that easily could be anything but a fad. A science cannot change its system concepts overnight, for precisely the same reason that an individual can't do the same thing.

A system concept is an attitude, an understanding, a world view. It's a sense of orderliness and coherence that we see in a body of principles and generalizations. It lives in an individual. It not only forms out of coalescing principles, but it determines which principles belong in the system and which do not. The process is one of assimilation and accommodation, simultaneous mutual adjustment between levels.

Acceptance of control theory requires a change in the beliefs of life scientists at the level of system concepts. System concepts bring order into principles; principles bring order into methods; methods bring order into symbolic representations; symbolic representations bring order into all lower levels of observation. Reorganizing a system concept therefore requires reorganizing everything else. The very way the world looks to us changes when a system concept changes. In fact, the system concept cannot change first. The whole system must reorganize at once. Newcomers to control theory do not all learn it the same way. One part of it is immediately clear to some, other parts to others. What we understand in one area of knowledge causes problems with what we thought we understood in other areas. Even in a willing individual, this reorganization can't take place overnight. It requires years. It requires changes at levels where we all find voluntary change difficult, mysterious, or even impossible.

In the life sciences, there is a widely-accepted system concept of what an organism is. When a scientist speaks from the viewpoint of this system concept, we can recognize it easily, although it's not easy to put into words. The words and descriptions we *can* find are only signposts pointing to the system concept. There's a dispassionate aspect to it, a distancing. There's an avoidance of empathy. There's a kind of sternness, an overcoming of natural sympathies, a pride in being immune to the weakness and sentimentality of the layman's view. There's a picture of an organism as a natural object, a bag of chemicals, a preparation of irritable tissue. The word most often used to symbolize this complex structure of attitudes is "objectivity." But objectivity is only evidence of the system concept: the system concept itself is a point of view from which all the rest, from principles on down, hang together and make sense. The system concept is the understanding of living systems that makes objectivity seem appropriate.

The control theory that you will find in this book is a collection of principles, methods, symbol systems, relationships, and observations of more detailed kinds. If I had it to do over again (and if I were a different and smarter person with a better education and a different way of growing up, and understood what I understand now), I would not persist so long in arguing at these levels. I would spend much more time trying to understand and express the difference in system concepts that separates control theory from all conventional theories. Control theory was never the only ingredient in this alternate view of organisms; I only made it seem that way. Even calling it "control theory" is an example of synecdoche, in the sense of referring to something by naming only one of its attributes.

Another important ingredient of this system concept is a view of what constitutes understanding of a system. I do not accept that statistical studies of behavior ever give us understanding of human behavior or human functioning. All they do is overwhelm us with random and unreliable facts. To understand a system, we must be able to see that it *must*, because of its inner nature, behave as we see it behaving. Its properties must grow out of its inner organization; its behavior must arise from its properties.

This principle is connected to a principle of explanation (or so, from my system concept, it seems). The only kind of explanation I can believe is one that demonstrates the principles that are proposed. It's not enough to say that a block diagram represents a system. One has to show that, in fact, a system organized in that way *must* behave in a particular way. If you can't deduce how a system would work from the explanation, then you don't have an explanation. The best way to prove that the explanation actually explains something is to cast it as a working simulation, turn it on, and let it operate by the rules you have put in it. If you can't do that, then you don't have a model *or* an explanation. All you have is more or less persuasive rhetoric.

I did not come down from a mountain with these principles tucked under my arm. They grew out of my training and my occupations, out of my successes and failures, out of my listening to others who were trying to solve the same problems. As the principles changed shape, the system concepts changed shape; as the system concepts changed shape, the principles became clearer. I still don't know how to express very clearly the system concept that contains control theory; now that I know this is needed, I will begin trying to do so. But I have learned that I don't have to provide understanding for others at this level. At best I can make it a little easier. But I don't really have to do it at all.

The reason I don't have to teach system concepts (aside from the fact that I can't) is that people can learn them on their own. The principles and methods of control theory can be taught; once a person grasps them beyond a certain point, they teach themselves. But those principles, everyone discovers, have implications that clash with the rest of one's knowledge of human behavior. Once one has understood an explanation of any one behavior from the standpoint of control theory, other explanations suddenly look different—more evasive and rhetorical, more conjectural, even wrong.

Understanding control theory just as a collection of logical and mathematical manipulations or as a diagram of relationships is relatively easy; those things can be taught to 30 people at once. You can learn those things from this book, reading it from either end. But grasping all the implications of control theory at the higher levels of understanding is hard, and can be done only within one individual at a time. Every control theorist I know has put a great deal at risk in the process of learning this subject. Most of them have suffered the embarrassment of seeing a former belief as foolish or naive, of realizing that they had been uncritical or even gullible. I know of none, however, who would go back to what they believed before, although there is nothing to prevent their doing so —and every encouragement from their colleagues to do just that.

Members of The Control Systems Group conceived the publication of this collection, organized it, and labored to make it real. I am profoundly grateful to them. Seeing all these papers brought together has, unexpectedly, shown me that a phase of my life is over. This book and the fact that I had so little to do with creating it have convinced me that I can stop writing that article over and over. The basic ideas are in good hands; I can let go of them now. Now, perhaps, I can try to remember what was supposed to come next.

William T. Powers
Northbrook, Illinois
July 1989

A Note on the Text

In resetting the text and redrawing the figures of papers included in this volume, I have kept alterations to a minimum (mainly silent corrections of obvious typographical errors and inconsistencies; in particular, the Editor's notes in "Applied Epistemology" were *not* added by me). The originals must remain as ultimate touchstones, and I alone am responsible for both intended and unintended differences between them and their reproductions herein.

My labors have been eased very significantly by William D. Williams, who arranged to have several of the papers put into computer-readable format, and by my wife Pat, who not only redrew some of the figures, but did so using software she had written herself. I am also grateful for aid from several CSG members and sympathizers, particularly those who helped with bibliographic work, as noted on page 295.

Special thanks are due to copyright holders of the papers reprinted with their generous permission.

Gregory Williams
Gravel Switch, Kentucky
July 1989

[1960, with R.K. Clark and R.L. McFarland]

A General Feedback Theory of Human Behavior: Part I

Introduction

In this paper we introduce a conceptual model of human behavior, based on some of the fundamental considerations of feedback theory and leading to a generalized theory of behavior. About six years of development lie behind what is presented here, so obviously we cannot explore in this one paper all the ramifications and applications of this theoretical structure which have occurred to us during this period. What we intend to do here is simply to present the theory as concisely as possible, so as to provide a basic paper in the literature to which we can refer when discussing experiments and further theoretical considerations in other papers.

The concepts presented in this paper represent a synthesis of many ideas, some of which have been in print for many years. Indeed, the literature of psychology alone, if interpreted in the light of what is known about feedback control systems, could be used to form the basis for our theory. Our approach did not begin from a psychological orientation but from the physical and mathematical, because the first two authors are physicists, who only after several years of work on this model, began to acquire a more thorough acquaintance with the work of psychologists. Thus, we find it most natural to develop the theoretical model first, before attempting to outline the applications of this model in language appropriate to psychology.

Reproduced with permission of publisher from: Powers, W.T., Clark, R.K., & McFarland, R.L. A general feedback theory of human behavior: Part I. *Perceptual and Motor Skills*, 1960, 11, 71-88.

2 Living Control Systems

At present we will present in Table 1 just 12 of the references in the literature which have given us key ideas and which have provided us with the necessary conceptual techniques. In later papers we will discuss the contributions of the psychological works mentioned here as well as many others, treating the major theorists and experimentalists in what we hope will be a thorough and orderly manner.

We strongly advise the reader who has something more than passing acquaintance with feedback *not* to skip over the initial parts of this paper in which we develop some of the basic feedback concepts. We have split up the generalized feedback system somewhat differently than is customary, and in our discussion of the operation of this type of system we will be introducing terminology to be used extensively later on in the paper. Furthermore, we have often found that some of our hearers have previously developed misconceptions about how feedback systems operate, which circumstance has led to pointless arguments about the properties of control systems. Before challenging our statements about how the generalized feedback control system operates, consult a servomechanisms engineer!

Fundamental Definitions

We will often employ the term "system" in this paper. Much work has been done on general systems theory, but we have found that for our purposes we have needed to formulate our own concepts, for convenience in discussing later ideas.

A system, as we use the term, is a collection of functions (not, as is often proposed, a collection of variables). A function is a relationship among several variables, and a variable is a combination of two classes of percept. Thus, to define "system," we start by defining "percept."

A *percept* is the basic unit of experience. It is that "bit" of perception which is self-evident to us, like the intensity of a light, or the taste of salt. In Part II of this paper we will give another definition which relies less on the subjective sympathy of the reader.

A *variable* is always a combination of two classes of per-

A General Feedback Theory of Human Behavior: Part I

Table 1. References Cited

1. ASHBY, W.R. *Design for a brain.* New York: Wiley, 1952.

See particularly paragraphs 1/1 through 1/6; note defects in 2/3 and 2/4; note that 2/7 implies strictly a transient-response study. Compare his "primary operation" with our "test of significant variable." Also see 3/11 for lucid discussion of feedback loops and lack thereof in most psychological experiments (Skinner's conditionally excepted).

2. FRANK, L.K., HUTCHINSON, G.E., LIVINGSTON, W.K., MCCULLOCH, W.S., & WIENER, N. Teleological mechanisms. *Ann. N.Y. Acad. Sci.*, 1948, 50, 187-278.

3. FULTON, J.F. *Physiology of the nervous system.* New York: Oxford Univer. Press, 1949.

Compare "Cerebral cortex: architecture, intracortical connections, motor projections," by Lorente de No. Pp. 288-330. See especially the diagram on page 307 for connections suggestive of upgoing feedback signals (a and a'), outgoing output signals, and imagination connections (S_3, S_7, S_5). Of course far too few connections are shown to perform any complex functions. This geometry is typical of most of the cortex.

4. HEBB, D.O. *Brain mechanisms and consciousness.* Springfield: Thomas, 1954.

5. HEBB, D.O. *The organization of behavior: a neuropsychological theory.* New York: Wiley, 1949.

"Phase sequence" and "cell assembly" are primitive feedback concepts. Many good examples of various orders of feedback control actions.

6. HICK, W.E., & BATES, J.A.V. *The human operator of control mechanisms.* (Monogr. No. 17-204) London: Ministry of Supply, 1950.

7. KORN, G.A., & KORN, T.M. *Electronic analogue computers.* New York: McGraw-Hill, 1952.

See pp. 4-10 for discussion of signal function, block diagrams. Note that the fact that the variables are *identified* as voltages has no bearing on the relationships discussed concerning their *magnitudes*.

8. KRENDEL, E.S., & GEORGE, H.B. *Interim report on human frequency response studies.* Wright-Patterson AFB, Ohio: Wright Air Development Center, Air Research and Development Command, USAF, 1954. (WADC Tech. Rep: 54-370)

A good example of what we are *not* trying to accomplish.

4 Living Control Systems

Table 1. (cont.)

9. SHANNON, C., & WEAVER, W. *The mathematical theory of communication.* Urbana: Univ. of Illinois Press, 1949.

The start of present-day "information theory."

10. SOROKA, W.W. *Analogue methods in computation and simulation.* New York: McGraw-Hill, 1954.

See Preface: rest of book is useful as demonstration that physical form of analogue is completely irrelevant to "behavior"; only relationships among magnitudes of variables are of interest for functional analysis of a system.

11. TRUXALL, J.G. *Control system synthesis.* New York: McGraw-Hill, 1955.

See particularly Ch. 2, "Signal Flow Diagrams and Feedback Theory." Note that roles of arrows in diagrams correspond to boxes in this paper, and nodes correspond to our arrows. Both representations are commonly used.

12. WIENER, N. *Cybernetics.* New York: Wiley, 1948.

See diagram on p. 121: the arrow labeled "input" is our *reference level*: this is thus conceived of as a system with *internal* loops. If X is taken to be our R, and "Multiplies Operator," the environment, the equations following describe our system for any one order of control. See also diagrams on p. 132.

---

cept. One class contains percepts which *do not vary*; by these percepts we keep track of the "identity" of the variable. The other class contains percepts which do change; these percepts carry the information about the "magnitude" of the variable. "Magnitude" is used here in its most general sense, including the meanings of "intensity," "size," or any other word for the general class of variable attributes.

A *function* is the direct relationship between any two or more variables. We shall uniformly imply by this term a *stable* relationship, which does not alter its form over reasonable periods of time. Since the variables we shall be talking about are assumed to correspond to physical events, we will always assume that whatever functional relationship is seen among variables is imposed by the operation of some physical "device," such as a neural network or a muscle or a chemical reaction. We shall sometimes represent these functions as mathematical expressions, in which case they are to be taken as idealized representations of some physically-occurring relationship.

A *system* is a set of functions interrelated in a special way. Given a set of variables and the physical devices which relate them in pairs or larger groups, we can define the environment of the system as all those variables and functions not included within the set chosen as our system. Within the defined system, in order for just one system to be under discussion, one must be able to trace relationships through the system (variable, function, variable, function, variable...) in a connected way such that no chain of relationships is independent of all the others within the system except for effects transmitted through an environmental loop. If the only relationship between two such chains of functions is through an environmental intermediary, then we would count two systems, not one.

The *input boundary* of a system we will define for the present purposes as the set of all functions which relate environmental variables to system variables *in a unidirectional fashion*; environmental variables affect, through some physical device, a system variable, but the device does not work backward.

The *output boundary* of the system will consist of all system functions which relate system variables to environmental variables, operating unidirectionally in the outward direction.

If any bi-directional function exists at the boundary, we would represent it twice, once as a unidirectional input function and again as a unidirectional output function.

All functions within the system will be treated as above; thus, we will be dealing strictly with unidirectional functions which may be *described* mathematically as working in either direction, but which in actuality operate in one direction only. Thus, for any function in the system we can define a variable or set of variables as the input to the function and a second set as the output from the function. We will often refer to such sets of variables as a single variable.

Finally, when we speak of variables we will be referring exclusively to the magnitude of the variable; its identity is incidental. In other words we are concerned only with information flow, and not with the means by which the information is transmitted nor the physical form in which it is transmitted.

Thus, we conceive of the whole system as basically an analogue, not a digital device. Digital functions can, of course, be constructed of such analogue functions. These considerations are not basic to our theory, but might explain some of our biases.

The Basic Feedback Control System

There are two major classes of feedback in common knowledge. One is the type which is wholly internal to a system, involving closed loops which do not cross the input or output boundaries of the system, and the other is the type in which the feedback path exits through the output boundary, passes through the environment (with attendant modification of the information) and reenters at the input boundary, the rest of the loop being completed within the system. Both types of feedback can exist simultaneously, but only the external type is unequivocally perceivable as a feedback loop by an external observer. The behavior of any system with internal feedback could be simulated exactly by another system with no internal loops, so such internal loops cannot be firmly identified by external observations.

We will be primarily concerned with externally connected feedback loops. Since we will be attempting to build a model of human behavior, we will regularly assume, unless special circumstances dictate otherwise, that the sense of the feedback is *negative*; this is, indeed, necessary if a feedback *control system* is to exist. The meaning of the term "negative feedback" will become apparent as we discuss the operation of the general control system.

The general control system consists of three functions plus an environment function, and five variables. We will discuss these in order from the input boundary, through the system to the output boundary, and through the environment back to the input boundary.

The input boundary consists of a function we call the Feedback Function, abbreviated F in equations. The environmental variable which is the input to this function we call v_e (which may represent, remember, many variables). The output variables of this function we call the *feedback-signal*, "*f*," reserving, as we shall do consistently, the term "signal" for variables inside the system. The feedback-signal is some function of v_e, the form of the function being determined by the properties of the input device. Mathematically, the relationship would be written

$$f = F(v_e). \qquad [1]$$

The next function is the Comparator Function (C), which receives both the feedback-signal *f* and a *reference-signal*, symbolized as "*r*." The Comparator Function subtracts *f* from *r* and its output signal is called the *error-signal*, "*e*," representing the discrepancy between *f* and *r*.

The function at the output boundary we call the Output Function, (O), which receives the error-signal as its input-signal and produces the *output-signal* (or variable), "*o*." This would be written

$$o = O(e) = O(r - f). \qquad [2]$$

The Comparator Function is often only implicit in the operation of the Output Function, some devices being capable of responding directly to the difference between two input-signals. For clarity we shall usually speak of the Comparator as a separate function and the error-signal, *e*, as a real signal inside the system.

The output variable *o* is the input variable to the Environment Function, (E), which in turn produces as an output variable (or set of variables) v_e, the input to the system. Thus, the loop is completed: see Fig. 1. We would write

$$v_e = E(o). \qquad [3]$$

For this system to be a control system, it is necessary that for any error-signal, the operation of all the various functions be such as to tend to bring *f* closer to *r* (in other words, to reduce the magnitude of the error signal). This is exactly what is meant by "negative feedback." If the environment offers no resistance at all to the output, so that *o* is capable of altering v_e

8 *Living Control Systems*

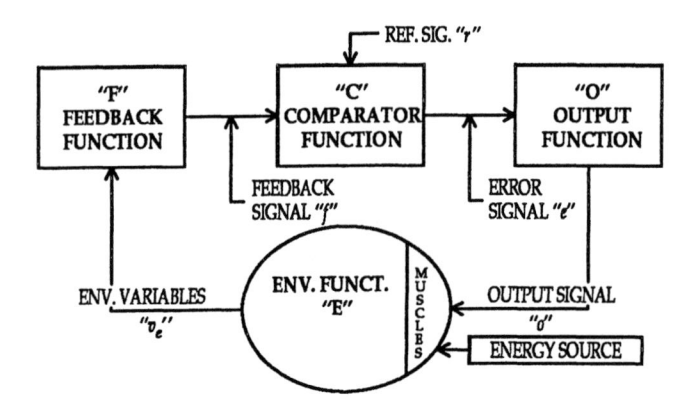

```
                    ┌── REF. SIG. "r"
                    ↓
  "F"              "C"              "O"
FEEDBACK  ──→  COMPARATOR  ──→   OUTPUT
FUNCTION        FUNCTION         FUNCTION
   ↑         FEEDBACK          ERROR
            SIGNAL "f"        SIGNAL "e"

ENV. VARIABLES   ENV. FUNCT.  M
   "v_e"            "E"       U   OUTPUT SIGNAL
                              S       "o"
                              C   ┌──────────────┐
                              L   │ ENERGY SOURCE │
                              B   └──────────────┘
                              S
```

Fig. 1. Feedback control system, general form

to any desired extent, then the system will come to equilibrium with the feedback-signal equal to the reference-signal. If the reference-signal is altered by some (unnamed) agency, the system will automatically respond to the ensuing error-signal by bringing f to the same (new) magnitude as r, thus erasing the error-signal and simultaneously reducing the output of the system to zero. For a system in this kind of environment, it can be shown that under all conditions within the operating range of the various functions, the feedback-signal will be caused by the actions of the system to "track" a slowly changing reference-signal. Thus, the reference-signal is the obvious means by which the system can be controlled.

In an environment which resists the output efforts of the system, or which introduces arbitrary disturbances into v_e, the system will still come to equilibrium, but an error-signal of non-zero magnitude will exist at equilibrium; this error-signal (or the discrepancy between f and r) will be just sufficient to maintain the output function at the right level of activity to keep equilibrium. In a reasonably efficient feedback control system, the error will be only a small fraction of the total magnitude of the reference-signal; the feedback-signal will still be maintained to a reasonable approximation "at the reference-level." Only when environmental disturbances cause some signal in the system to exceed the level its associated devices can handle would we expect to find any appreciable discrepancy between f and r.

For the benefit of the reader familiar with transfer-function studies presently being conducted by many workers, we should mention that we are concerned here only with the steady-state relationships in these control systems. We view any such system, therefore, on a time-scale on which transient disturbances occupy so little time that we can neglect them. For some human systems, this may mean that we pay no attention to intervals smaller than 0.1 sec., and for others, that we ignore all events lasting less than several seconds, minutes, or even days. By limiting ourselves to consideration of quasi-static equilibrium, we have found that the over-all organization of a complex system is much easier to conceive. This does not imply that the system is motionless, but only that all error-signals remain small, the feedback-signals normally being maintained at whatever value the reference-signal may have for the time being. A system in which all error-signals are comparatively minute could still be engaged in violent activity, as various reference-signals are altered to cope with a changing environment.

A final word on this basic feedback unit. We are going to use it as the building-block (with some modifications) of a complex many-leveled system. If we were faced with the task of designing such a system that would actually be overall-stable, not oscillating wildly or locking itself up in internal conflicts, we would give up right here. Fortunately, we are not concerned with design criteria, for the human system we deal with is normally very stable, with no crippling conflicts and no obvious uncontrolled oscillations going on. Thus, questions of stability criteria, non-linearities, limits, and the like do not concern us in our basic attempt to construct a man-like system. We assume that the various functions have forms, including transient response terms, which result in stability, so that by leaving the details of the functions unspecified, we have by definition a stable system. Later on, when the model is completed, we can consider a few of the pathological conditions that might correspond to conflict among feedback systems and various forms of instability.

Aggregates of Feedback Control Systems

Let us consider a collection of functions in an extensive system (which may in some cases prove to be more than one system). As we have already noted, some of these functions will be members of the input boundary, others of the output boundary, imposing relationships between system and environmental variables, in one direction or the other.

Some of the boundary functions will be found to form feedback control systems (in pairs, one input system and one output system) with perhaps some intermediate function within the total system. All such boundary feedback systems we will classify as first-order systems. In the human being, these boundary systems correspond largely to what have been unfortunately labelled as the "spinal reflexes." The spinal reflex systems are fairly efficient control systems having proprioceptive inputs and motor outputs and receiving reference-signals both in the output function (muscle-bundle) and in a comparator function (ventral horn cells). Indeed, these first-order systems almost monopolize the output facilities of the organism. There are input functions, however, which are not part of these control-loops.

Idealizing from this neurological hint, we will restrict our model so that *all* its output boundary functions belong to first-order control systems, and none are controlled directly and exclusively by "higher" systems. We allow some input functions to generate signals within the system which are not part of first-order control-loops.

In the human systems, it is the rule that many first-order systems affect the same variables in the local environment and thus affect each others' input variables v_e. It will be common, then, that many first-order systems will act as environmental disturbances on the inputs of other first-order systems. These disturbances will be corrected, or at least resisted, by each local system, and chaos will obviously result if reference-signals are not properly coordinated.

We can now select out of all the remaining functions in our system those which form *second-order* control systems to perform this coordination. These control systems will receive not only the output-signals from some of the "unused" first-order input functions, but will also receive as inputs the same variables which serve as feedback-signals in the first-order systems [in the human system, it is well known that the proprioception feedback-signals in the first-order spinal loops (and peripheral nerves in the cranium) divide, one branch going to more central systems].

Thus, if we wished we could now define a second-order input and output boundary; crossing the input boundary will be all or most of the signals generated by first-order feedback functions, whether involved in the first-order loops or not, and crossing the output boundary will be a set of output-signals which enter the first-order systems. These signals cannot be considered as adding to the outputs of the first-order systems, because feedback systems tend to go into violent conflict if their outputs are tied together, thus inactivating those systems (the theory of conflict will be discussed later). The only feasible control-point is the reference-signals of the lower-order systems; therefore, in our model we identify (for the time being) the output-signals of second-order systems with the reference-signals of first-order systems. To put it graphically, the output of a second-order system is not a muscular force, but a goal toward which first-order systems automatically adjust their input-signals (proprioceptive sensations). Thus, the second-order system acts, so to speak, by specifying for the first-order system the kind of sensation it is to seek; the first-order system adjusts its output until its input-signals match as closely as possible, in the given environment, the "example" given by the reference-signal, thus (quite incidentally) producing environmental effects which an external observer could see.

This viewpoint is extremely important to understand: in all the feedback systems we will discuss, it is of no concern at all to the feedback system what actual effects are produced in the environment. The system reacts only to the signals injected into it by its feedback function, and for any one system nothing else exists. Even when we speak of systems which deal in human interrelationships, these complex systems not only do not "care" about what is actually going on in the "real" environment, they cannot even know what is going on "out there." They perform the sole function of bringing their feedback-signals, the only reality they can perceive, to some

12 Living Control Systems

reference-level, the only goal they know. If we were discussing servomechanisms, such anthropomorphisms would be unnecessary, but when we are talking of the very systems in which we live, now and always, which we must employ even to think, anthropomorphism is an essential ingredient of understanding.

It is evident now that we could go on defining successively higher orders of control until we had exhausted our collection of functions. We would then find all the sub-systems, each a feedback control system, arranged in a hierarchy (or many overlapping hierarchies) in which a system of any one order perceives an environment made up of the feedback-signals of the systems in the next lower order, and which acts to change that environment by producing output-signals which are the reference-signals of the same lower-order systems. This structure is exactly the basic organization of our model. A model of this type could be constructed (ignoring practical difficulties) which would reproduce any kind of human behavior that did not involve changing the form of any functions or adding new systems to the structure: the model thus far is intended as a model of those human systems which produce learned behavior, *after* learning has taken place. This model, being built entirely of feedback control systems, is inherently capable of maintaining dynamic equilibrium (error-signals small, but not necessarily a physically static system) in the presence of a wide variety of environments, both familiar and strange. It is "adaptive" to the extent that it can cope with a large variety of new environmental *configurations*, but it cannot do a thing about an environment which changes its *properties* (summed up as the E-function in Fig. 1). We still lack something to account for non-rote learning, for that requires altering the *structure* of the system, not merely its information content.

The Negentropy System

We borrow the term "negentropy" from information theorists to refer to the process of decreasing entropy in a local system (at the expense, of course, of increasing entropy elsewhere), which process has been identified by some with an increase of organization within a system. We conceive of the central nervous system as being a collection of neurones forming a complex and largely random network, which can have its effective structure altered by activating and inactivating connections within the net to produce networks with semi-permanent and well-defined functions, which to human beings would appear less random.

The processes which alter the connections within the basic bed of "uncommitted neurones" (McCulloch's term) to form the various orders of feedback control must themselves represent the working of a system which is *not* the result of learning, but which is present and active from birth or before. This system may be physically indistinguishable from the resulting learned systems (perhaps it is implicit in the "random" connections in the "unorganized" neurones), but it is functionally quite different. Its output must be complicated and must extend throughout the CNS, because systems which have been learned are apparently subject to further modifications or additions. Rather than attempt to postulate what the nature of this output must be, we will define it simply in terms of what it must do.

The output of the *N*-system, we hypothesize, results in the following kinds of events. (1) Uncommitted neurones in physically suitable regions become tentatively organized to process a number of feedback signals from the highest existing order of control (which in the beginning may be first order). (2) Other uncommitted neurones likewise undergo tentative organizations which generate signals serving as reference-signals for the next lower order of system. (3) These tentative organizations of input and output can occur at a variable rate. (4) When a particular organization has occurred often enough[1] within a collection of uncommitted neurones, the organization tends to persist, and the input and output functions of a new order of control system have been formed (as Hebb and others have suggested).

Thus, we have *identified* the output variable of the *N*-system as "the processes which alter organization in uncommitted neurones" (as well as in existing systems). The *magnitude* of

---

[1] "Often enough" means one or more times.

this variable we postulate to be measured by the rate at which new organizations are formed one after the other.

The changing organizations occurring in potential output functions will result in a continuous alteration of the reference-signals in the momentary highest-order systems; this results in observable trial-and-error behavior, which shows some organization owing to the existing hierarchy. The continuing reorganization occurring in the new input function does not have such externally-observable results, but is subjectively recognized as a kind of trial-and-error effort to perceive new patterns, a common experience in a learning situation which includes what we experience as tentative formulation of hypotheses. The "hypotheses" here should be thought of as tentative definitions of new variables, which may or may not prove to repeat themselves in experience, depending on the organization of lower-order perceptual functions and the properties and nature of the environment.

The input variables which affect the input boundary of the N-systems we call "intrinsic signals"; we suppose these to be a set of sensory signals which are measures of a set of physiological states, including but not necessarily limited to the ones commonly associated with the "drives." When these variables are each at some certain critical level, the organism is operating optimally, as far as the N-system is concerned. There may be many effects, such as those due to radiation damage, which are deleterious to the organism, but which are not directly represented by intrinsic signals.

The N-system we assume to be a feedback control system which is organized to maintain the intrinsic signals at particular reference-levels. These reference-levels may be set by neural signals (as, perhaps, for sex or hunger signals) or they may be determined by the physical properties of the N-system functions. In either case, the reference-"signals" must be genetically determined, not determined by experience, for the N-system must be a complete control system (which implies reference-signals in existence) before any learned system can be developed. When all intrinsic reference-levels are satisfied by their respective signals, we say the organism is in its *intrinsic state*.

The overall operation of the N-system is thus very easy to describe (see Fig. 2). If some event occurs which makes one or more of the intrinsic signals depart from its reference-level, the N-system produces an output-signal proportional (as a first approximation) to the error. Since the output-signal has been defined as a rate of reorganization of neural networks, the net result is to establish a certain rate of attempting to learn. We would say "rate of learning" except that whether or not anything can be learned by reorganization depends to an important degree on the nature of the environment. If the reader will keep in mind this hedge, we will after all use the more convenient expression "rate of learning."

Simply put, the rate of learning is approximately proportional to the intrinsic error-signal, and this is a fundamental property of the human organism.

A *particular* organization will become a stable learned feedback system not because there is anything that "tells" the system to stop reorganizing, but because the lower-order systems and the environment are such that this particular organization

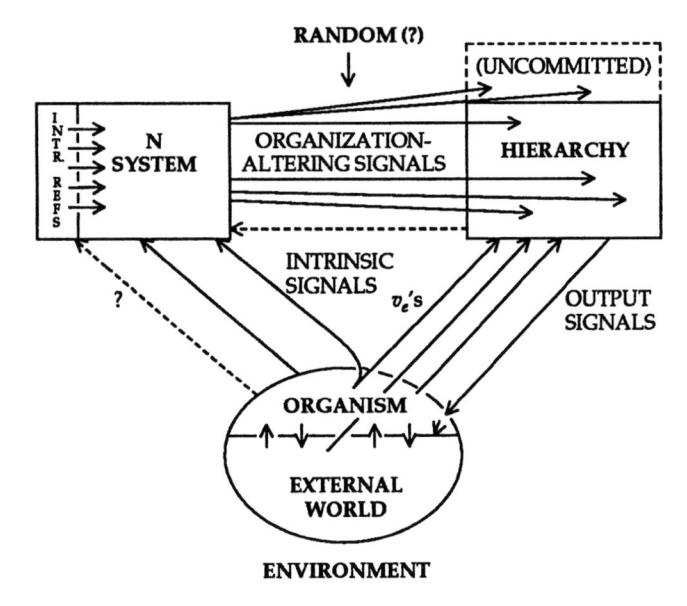

Fig. 2. Overall organization in model

produces behavior which results in a lessening of the intrinsic error, thus slowing or halting the reorganization process. If the same organization proves to have an intrinsic-error-reducing effect several times, then reorganization will stop with the new higher-order system in approximately the same form several times, and we suppose that this will cause the organization to tend to persist,[2] or even to become a semi-permanent part of the hierarchy of learned systems. This kind of learning has many evolutionary advantages; for one, a new system will not be fixed for every chance arrangement of the environment, but only for situations which tend to repeat. Another advantage is that while reorganization will stop with the new system in *approximately* the same form as before, there will tend to be differences in detail, so that the "noise level" is reduced, much as one eliminates irrelevant variations from planetary photographs by superimposing many negatives to form a composite print.

Modifications of the Basic Feedback Unit

Our model so far has many properties like those of human beings, but we are lacking several important ingredients (at least!). The model has no memory for past experiences, it cannot use past information in present actions, and it is incapable of imagining (which we define as the ability to perceive sensory events generated internally rather than generated by present-time interactions at the input boundary of the whole system). As we consider them, memory and imagination are fundamentally related.

To see how we propose to introduce the function of memory, refer to Fig. 3. A new block has been added labelled "R," which stands for the *recording function*. We assume that there

---

[2]Because this form is, therefore, approximately adequate for control of the existing environment and hence will be changed further but little. This does not imply or deny a frequency theory of learning, for each organization that exists when learning ceases has been "learned," whether or not it is learned completely and whether or not it is an appropriate form. In this sense, learning is always complete, but perhaps does not match what E has in mind as the "proper" final organization.

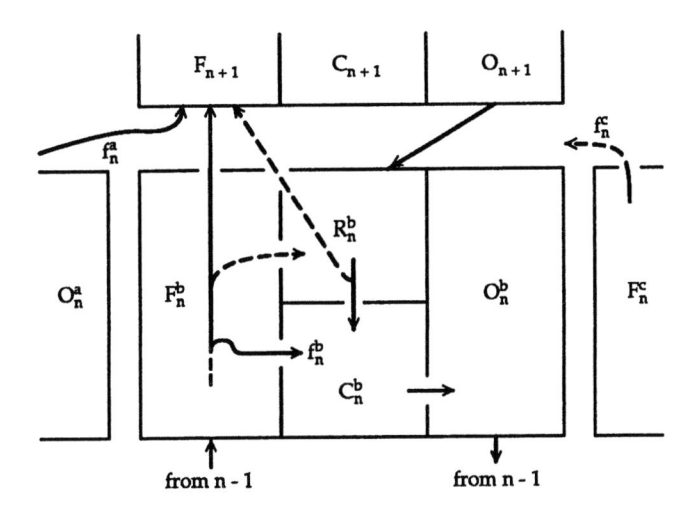

Fig. 3. Relationships among orders

is a recording function associated with *every individual feedback subsystem* (associated functionally, not necessarily in space).

This recording function has an input which is the same feedback-signal used in the local feedback loop and sent to higher-order systems. The function R receives this signal and by some means neither we nor anyone else understands, records the information carried by it. The result is a set of recordings which may be permanent or which might have some finite half-life. (There is no present way to tell whether forgetting is due to fading of the recordings or to failure of the recovery apparatus.)

The recording function has the further property that when it is selectively stimulated by a signal external to the local system, it will produce a signal which is a facsimile of the signal that was recorded. This reproduced signal carries the same information, or some significant portion of it, that the original feedback signal carried. To all intents, it is a sensory signal, but one arising from a past event rather than a present one. Current experiments in brain stimulation tend strongly to support this view of memory.

It will be noticed that the signal from a higher-order system in Fig. 3 no longer serves directly as the reference-signal for the pictured system. Rather, the higher-order output-signal stimulates a memory-trace in R, which in turn produces a signal that is used as a reference-signal in the associated subsystem. Thus, the reference-signals which control a given feedback unit are examples of its own past sensory signals, and one could now express the task of the control system as being that of reproducing in present-time experience some previously-experienced perceptual field, or portion thereof. To some degree new perceptual fields could be demanded and brought about by stimulation of combinations of memory-traces. Rote learning could occur in the form of new recordings and hence an enlarged repertoire of reference-signals.

The process of selecting a memory-trace and stimulating it might be a function of R, or it might result from some property of higher-order Output Functions. We have not tried to specify the processes involved any further than our statements about what we assume to happen. In either case, the overall effect is that higher-order Output Functions act by stimulating memory-traces in lower-order recording functions.

We have come to associate perception with feedback-signals, and specifically *not* with output-signals. A moment's introspection will convince the reader that he *never* perceives an output-signal in his own system. Even muscular forces are perceived as proprioceptive sensations. Thus, if the objects of perception must all be the signals f, our model still cannot remember! It cannot, that is, perceive signals arising from its memory-traces, because as we have drawn it so far, the reference-signal that is the remembered feedback-signal enters the Comparator Function, which is associated with O, not F. We have a situation of some psychological interest wherein our model can reproduce a past experience without being able to perceive that experience.

The reference-signal carrying "imagined" information cannot be properly interpreted by the Feedback Function F of the associated system of the same order, at least not in general. This is best demonstrated by an example.

Suppose that F receives a single variable x and squares it to produce a new variable y:

$$y = x^2. \qquad [4]$$

If this new variable y were to appear at the input of F, a new variable y_1 would be generated, equal to y^2 (because the function always performs the same operation on its inputs). Thus, we would have

$$y_1 = y^2 = x^4. \qquad [5]$$

We see that the new variable y_1 represents x^4, which is not the same "interpretation" given to other lower-order signals received by F. Thus, the system could not act correctly with respect to such a twice-processed variable if it were set up to handle variables representing x^2.

It is true that certain functions will not introduce such a distortion if applied to their own output signals (e.g., if $y = x$, then no distortion will result from any number of reprocessings), but the general structure of the model cannot be made dependent on such special cases; the way the model is to handle the imagination information must work for *any* form of F.

If the reference-signal is indeed a reproduction of a past feedback-signal, then it bears the same relationships to lower-order signals as do present-time feedback-signals in the associated system. Therefore, in view of the previous paragraph, if the reference-signal were to enter a Feedback Function of the next higher order, it would always be interpreted properly, just as are the feedback-signals currently present. Consequently, we introduce into the hierarchy what we call the "imagination connection," shown in Fig. 3 as a dotted line splitting off from the reference-signal in one system and entering the Feedback Function of the controlling higher-order system.

This connection is shown dotted; its introduction must be qualified because of the effects of having this connection present.

Note that the higher-order system would find its feedback signal at the required reference-level solely on the basis of the imagination-signals from lower order, even though the lower-order signals might be quite far from the reference-levels in the lower-order systems. This could occur if the higher-order F received imagination-signals *in preference to* feedback-signals;

a condition like dreaming or fantasy would occur, in which every goal set for the lower-order systems appeared to be immediately satisfied—in imagination, of course. This might seem clearer if it is remembered that normally the higher-order system specifies a reference-signal which the lower-order system matches with its own feedback-signal; if the reference-signal substitutes for the feedback-signal, the "match" is automatically ensured.

The imagination-signal makes it possible for our system to perceive reproductions of past perceptual signals (that is, to remember as well as record), to plan an action "mentally" without actually performing it, to hallucinate, and as mentioned, to dream.

Obviously, the hierarchy could not perform very reliably in a real and sometimes dangerous environment if its actions were completely "short-circuited" by the imagination connections. Somehow this configuration must contribute more information to the perceptual field at some times, less at others. Under conditions of sensory deprivation, it apparently provides a great deal of information, while under conditions of, e.g., immediate danger (barring pathology) it contributes little. Everyone knows that the more thoroughly one wraps himself in perception of internal events—thoughts, memories, daydreams—the less sensitive he becomes to the present environment. There appears to be a kind of mixing control, which can be adjusted to full imagination (as when asleep) to full present-time perception. This might be a property of the Feedback Functions, corresponding to a shift in perceptual attention, or of the manner in which Output Functions stimulate lower-order recordings. We are open to suggestions.

The normal condition is probably one in which most information is present-time perceptual information, and small errors are filled in by the imagination connection—this would be a pro-survival property, in that it would allow the feedback systems to be very exact in their control-actions, while not tying them up over trivial discrepancies. The phenomenon of "filling in" small discrepancies is well-known under the label "closure."

Summary of Part I

What has been presented so far is a model, a collection of functions which handle signals, arranged into a hierarchical structure and composed of elementary feedback control systems of the external-loop type. For the feedback systems of any one order of control, the environment consists of a set of feedback-signals, the same ones used in the control-loops of the next lower order; this environment is controlled by means of signals sent into the lower-order recording functions.

This set of systems is controlled by signals from higher orders or from random reorganizations of potential higher-order Output Functions in the bed of uncommitted neurones; such control signals stimulate the recording functions in the controlled system so as to give rise to reference-signals, reproductions of past feedback-signals produced by the local Feedback Functions.

The rate at which reorganizations take place in this hierarchy is proportional to the degree of intrinsic error existing in the N-system, which is a feedback control system of the external-loop type concerned with maintaining a set of intrinsic variables at their genetically-determined reference-levels; the function of the N-system is to maintain the organism in its intrinsic state, or as near to it as possible. The output action of the N-system is conceived of as essentially random.

While we have made occasional reference to psychological or neurological properties of human beings as a means of making certain points more acceptable, this portion of the paper has been primarily concerned with presenting the structure of our model, not its application to understanding human behavior. Part II will deal with the problem of translating from this functional scheme to terms appropriate to human beings. The two parts are (understandably) reversed from the order in which this whole picture was developed.

The operation of this model can be summed up perhaps more clearly in plain language. A system at a given order has goals given to it by higher-order systems. These goals are in the form of perceptual images of past experiences or combinations of past experiences. The system acts to make its present perceptual field match the goal-field as nearly as possible. It does not act directly on the external world, but on the only environment with which it is in immediate contact, the set of next-lower-order systems. Its action is that of selecting and stimulating goals for lower-order systems; it is capable of perceiving the signals (either feedback- or reference-) resulting from its selection, so a set of lower-order signals can be specified which, if achieved, would be interpreted by the system's own Feedback Function as the required magnitude of perceptual variable.

Only first-order systems act directly on the (non-CNS) environment.

Comments

To an external observer the behavior of this model could, in principle, be interpreted at many different levels, each quite correctly. This follows from the fact that the feedback-signals at a given order are variables which represent the collective behavior of some set of lower-order variables, and so forth down the chain of command, so that at each order we find the feedback-signals corresponding to variables abstracted farther and farther from the original raw sensory data and individual environmental events. Each order of system acts on the lower-order systems until it perceives its own kind of variable as being at the required reference-level. It will alter its outputs to the lower-order systems to counteract environmental events which have, via intermediate perceptual interpretations, a disturbing effect on the feedback-signal.

Thus, if one knew the kinds of transformations that characterized the transition from perception at one order to perception at the next order, he could observe the environment of the system under study and make parallel abstractions of his own; he could thus define nth-order variables in the environment of the other system, and watch how the other system interacted with those perhaps quite abstract variables. He could tell if those variables were actually under feedback control by the other system simply by applying forces to the environment which tended to alter those variables (but not inexorably, else no feedback action could occur) and watching to see which if any were maintained constant by the behavior of the system. He could tell whether he had abstracted correctly (to any desired probability of correctness) by applying all the different kinds of disturbance he could think of; if the variable were maintained constant or nearly so against all those disturbances, he could be fairly sure he had abstracted properly; that is, in the same way that the subject system's Feedback Functions abstract. By the same token, he could discover the reference-levels at which these variables are being maintained.

Given enough acquaintance with the system under study, the observer would see that the system is *always* maintaining *all* orders of perceptual variable at some momentary reference-level, by an active error-correcting process, except when its abilities are overwhelmed by superior forces in the environment. Even then, the higher-order systems will compensate by readjusting the reference-levels of lower-order systems, which might be seen as a drastic shift in the whole mode of behavior —from fighting to fleeing, perhaps. Whether fighting or fleeing, however, the lower-order systems would still be seen to control successfully patterns of movement, coordinate spatial relationships, produce vector forces, and so forth in a stable and a disturbance-resistant manner.

If a human being is indeed this sort of functional being, we can find out more about what is going on inside him if we can learn to understand the various classes of perceptual variable which are involved in his feedback control systems. The method of disturbing and testing, which we call the "test of the significant variable," is one method, and it is wholly scientific in its procedures, but fortunately we need not go through this tedious process to obtain every bit of information we are going to accumulate. Both the human subject and the investigator are presumably similar creatures, and the investigator can often find short-cuts by an introspective analysis of his own perceptual methods. This, of course, cannot be done in the sense that the investigator cannot perceive his own perceptual apparatus. He can, however, attempt to discover those variables which in his experience are *self-evident* classes, that is, which to his knowledge and belief are the forms in which he must perceive and always has perceived his universe. This approach is naturally subject to errors of idiosyncrasy, but the results, in the form of classes of variables which should be significant to other human control systems, can be subjected to the test of the significant variable, and false or inaccurate guesses eliminated.

To give the reader an advance notion of what we mean by a "self-evident class of variable," consider the referent of the term "sequence." This is one of the self-evident classes. We do not mean that everyone calls this part of his experience by the term "sequence," or even by any related term. That is part of verbal behavior. What we mean is that we think every human being can perceive the difference between experience A occurring before B, and B occurring before A, provided the limits of perception are not approached. He can set up a control system that is capable of reproducing a past sequence of simple events correctly, in the same order as originally. If he cannot do this, he cannot talk, he cannot reason, he cannot even detect the passage of time. If he did not perceive and control variables of sequence, he could not be sure of walking forward rather than backward, and although he might be able to recognize his telephone number visually, he could not dial it.

Furthermore, "sequence" is a unique category, qualitatively different from other categories. A simple sequence (the least element in a sequence of sequences) is perceived as an entity different from any of the individual static *configurations* of which the sequence is always composed. A sequence can be maintained even though the individual configurations used to produce it change. I can hum "Shave and a haircut, six bits," or I can drum out the rhythm on the table, or I can reproduce the rhythm by generating nine different sensory impressions in the right order (pauses and sounds): 0—&@%,——$^1/_2$ $^1/_4$. But I must always employ some set of static configurations, for that is another self-evident class of perception, and it is the *next-lower order* of perception.

Discussion of these categories of perception (which is sufficient to define categories or orders of control system) will occupy most of the second section of this paper.

[1960, with R.K. Clark and R.L. McFarland]
A General Feedback Theory of Human Behavior: Part II

Introduction

The model described in Part I is only a part of our general theory—the part which organizes our more general ideas about human behavior and human nature. To conceive of human organization as following that of our hierarchical array of FBCS (externally fed-back Feedback Control Systems) implies a certain attitude toward behavior, different in some important respects from traditional psychological viewpoints. Some of these differences we began with, but most of them took form only as we went back and forth between modifying our organizational model and observing people behaving.

One of the most puzzling, and in our opinion critical, aspects of human behavior is that behavior appears multiordinal. The same behavior can be described in a number of apparently equally-valid ways, from the particular to the general. Usually this representation of human behavior at varying levels of abstraction is put aside during a scientific study, and one particular level is chosen as the most interesting, or sometimes as the only "proper" one. But for us this multiordinality raised a critical question: is it due to the way in which behavior is observed, or is it somehow a significant property of the behaving system?

The answer we have arrived at is, "*Both.*" One must never forget that the person observing human behavior is a system

Reproduced with permission of publisher from: Powers, W.T., Clark, R.K., & McFarland, R.L. A general feedback theory of human behavior: Part II. *Perceptual and Motor Skills*, 1960, 11, 309-323.

like the one he is observing. If we accept that our model represents behavioral organization, particularly in the FBCS aspects, then it is *perception which gives form to behavior*. Behavior will make sense to E only if E knows *what perceptual variables the behavior is maintaining at some reference-level*. If an organism is producing behavior as a means of controlling a several-times-abstracted variable, then E has no hope of seeing order in this behavior unless he is capable of learning to select out of his experiences the relevant elementary sense-impressions and then can combine them in the same way that S is combining them to make a perceptual variable. If S and E are both FBCS, then even in a varying environment requiring widely-varying physical action, S will be able to maintain abstract variables at reference-levels, and E, if he resembles S, will be able to perceive that S is doing so.

The Human Hierarchy

We have developed definitions of six orders of control systems, giving the corresponding orders of perceptual variables names which represent classes of perception. These classes appear to human beings to be self-evident aspects of directly-perceived sensory fields which we call the "external world." Once one learns to perceive in these ways, the resulting impressions appear to objective, and one has the feeling of having "discovered" them, in an insightful way. Why it is that through learning one should develop just the six orders we postulate we cannot answer—perhaps the unseen external reality is so structured that we *must* learn to perceive in these ways in order to control our environments, perhaps our brains are so constructed that development of certain types of perceptual transformations is favored, or perhaps these orders of perception are peculiar to our culture, or even to the authors' microcosm! Leaving this problem for the future, we will propose our definitions of the six orders of perception and FBCS which we have been able to work out, and assume for the time being that all people are organized this way. Of course future experimentation will be specifically directed toward testing that idea.

Are adjuncts testable?

The classes of perceptual variables we will define bear the same relationship to each other as do the feedback-signals in the model of Part I. The higher are derived from sets of the lower, and at the same time contribute to still higher-order perceptual variables. Each order consists of a great many individual FBCS each of which controls its individual one-dimensional feedback-signal toward a reference-signal set by a higher-order system. The highest order of reference-signal is set by noise or by random action of the N-system. (See Part I.)

In more common terms, the *purpose* for controlling a given perceptual variable toward its reference-level is that of maintaining a higher-order variable at its reference-level. Higher-order perceptions are kept in their goal-states by *specifying* lower-order goal-perceptions; the higher-order system decides (no quotes) on a goal-perception for the lower-order system, but does not actually do anything to achieve it. Thus each goal-seeking system is autonomous to the extent that it must contain the circuitry for making its own feedback-signal approach its given reference-level and for recording its own store of potential reference-signals for later use: but each goal-seeking system is controlled to the extent that it does not choose which of its past experiences are to serve as goal-perceptions.

Complexity is not a factor in determining relative order, and neither is number of perceptual elements contributing to a given perception. *?* An nth-order variable can be exemplified by a set of lower-order variables (provided the observer has nth-order systems) but it belongs to a self-evidently different class of perception from the lower-order variables themselves. In order to perceive and control nth-order variables, one must simultaneously perceive and control lower-order variables (except for $n = 1$), but the reverse is not true. Eliminating higher-order perceptions leaves the lower; blocking the lower partially or totally eliminates the higher.

The preceding paragraphs indicate the kind of rules by which one can find a higher-order variable given a set of lower-order variables, or by which one can analyze a higher-order variable into lower-order variables which contribute to it. *All* these criteria must be met, and to understand our definitions properly, the reader must check his understanding against these "rules," hazy as they may at first appear.

In the following, S_n = a typical FBCS of order n; f_n = feedback-signal in S_n; r_n = reference-signal for S_n; F = Feedback Function of S_n (see Part I).

We have found a very simple demonstration which is probably the best way of clarifying the first four orders, and perhaps the fifth order as well. The equipment is cheap—two people, S and E.

First Order

First-order systems we identify almost exclusively with the spinal reflex loops, which are FBCS. These FBCS maintain proprioceptive feedback signals from very limited portions of the environment (tissue, tendons, etc.) at levels specified by the excitatory reference-signals descending the spine. Similar loops involve some cranial nerves. Many signals arising from sensory endings are not involved in S_1 (first-order control systems), but for convenience we generally refer to all primary sense signals as f_1, first-order feedback-signals.

Demonstration of First Order

S extends his arm in front of himself, with instructions to hold it steady, and E places his hand lightly on top of S's. E gives a sudden sharp downward push, and S's arm appears to rebound as if on a spring. An electromyograph verifies that this is an active innervated correction and not simply muscle elasticity. The initial position of S's arm makes no difference, and the initial muscle-tensions involved (as long as they are not zero) therefore make no difference to this response, thus showing that the reference-levels for the many systems can be adjusted and that the systems will correct their inputs toward any given reference-setting.

Second Order

Second-order systems S_2 derive their f_2 from sets of f_1. We call the class of all f_2 "elementary sensations," since they represent the initial grouping of the undifferentiated f_1 into elements with characteristic sensory patterns. In the kinesthetic modality, these would be made of signals representing muscle stretch, joint angle, tendon tension, and internal tissue pressure, which add up to the elementary sensation of effort and a kind of absolute sense of position (not relative limb position), like the pattern of signals one gets from clenching his fists. These elementary sensations, f_2, have recognizable patterns by which we identify them; for this reason we sometimes refer to f_2 as identity signals.

Demonstration of Second Order

E now instructs S to extend his hand as before, and E places his hand on top of S's. Now E tells S to swing his arm downward as rapidly as he can, as soon as he feels E push down. E's hand must begin in contact with S's to make the push as sharp and unexpected as possible.

Immediately after the push, the S_1 return the arm to its initial position, because they act *within the latent period* of S_2. Then, after the return swing is nearly completed, the S_2 react by resetting the r_1. The S_1 are then abruptly given new reference signals and accelerate the arm downward as requested. S cannot eliminate the return swing at the beginning of the response—if he could, he might be subject to instability.

Third Order

S_3 combine f_2 and/or f_1 to produce f_3, which we call "configuration" or "arrangement" signals. These represent any static combinations of sensations. At third order, many different arrangements give different signals in a given system, although a single S_3 will sense the same arrangement (the same magnitude of the signal f_3) for a number of different sets of f_2. Each third-order system can thus sense a limited range of arrangements of these f_2 which it senses, and ignores or fails to differentiate between arrangements of f_2 which do not yield different f_3 (the mechanism of "equipotentiality").

Hand-eye coordination involves, quite often, controlling an arrangement of visual objects toward some static reference-arrangement. In our Portable Demonstrator, the arrangement to be adjusted is the relative position of the S's index finger-tip and E's.

Demonstration of Third Order

E instructs S as in the second-order demonstration, but requesting that the movement be made sideways, and again making the initial press in the direction of motion. Now, however, E extends his other hand, holding out his index finger, so that S will have to move his arm about a foot to eighteen inches to touch E's index finger. S is instructed to extend his own index finger, and to swing his arm as quickly as possible after the push and align his finger with E's as *rapidly* and *accurately* as possible, so that the fingertips just touch. At the instant of the push, *E shifts his target finger 4 or 5 inches*, lowering, raising, or retracting it.

The first two orders of reaction remain visible, and at the end of S's rapid swing a third phase shows itself; S's finger comes nearly to a stop *near where E's finger initially was*, and then begins a much slower corrective movement quite different in nature from the first two actions. This third phase is the third-order reaction, showing a still-longer latent period. The second-order systems achieve their goal-states much more quickly than third-order systems; so quickly that under proper circumstances they actually have to wait for the next reference-level to be set by the controlling third-order system.

Fourth Order

S_4 convert sets of f_3 into f_4, which we postulate to represent *sequence*. That is, a given sequence of appearance of the f_3 (or f_2 or f_1) will yield a characteristic magnitude of f_4 in the S_4, and a different sequence may yield a different magnitude. This relationship holds, of course, only for the limited set of f_3 to which a given S_4 is sensitive. Fourth-order feedback functions F_4 must necessarily be rather complex devices, having short-term memory capabilities (as distinguished from the recording properties common to all systems); also they must be considerably slower than the F_3 for stability of control.

It is important to remember that a static feedback-signal at fourth order represents a *continuing* sequence, a constant shift of reference-levels r_3. (If the sequence ceases, has any of its lower-order elements modified, has its tempo changed, etc., the f_4 must change.)

Demonstration of Fourth Order

E instructs S to extend his index finger and track E's index finger as accurately as he can. E then moves his own finger in a circle 8 to 12 inches in diameter, gradually speeding it up until S is tracking smoothly (about one cycle per second). Without warning, E stops his finger dead still at some point in the circle. S continues to "track" for nearly half a second before being able to stop the independent sequence he has set up. His reaction time does not shorten significantly with practice. Since we know S is physically capable of arresting a motion much more quickly than this, the lag is due to the slowness of the fourth-order systems. Unfortunately we have not been able to think of an experiment in which the reactions of the first three orders are visible along with the fourth-order reaction. In most Ss, third-order responses can be observed just as S begins tracking; one sees a succession of jerky corrections, as the third-order systems attempt to correct one error in static configuration after another. This is soon supplanted by a more refined fourth-order response as S learns the appropriate sequence of movements.

If E, instead of merely stopping his hand, jerks it suddenly away, S will show a much faster reaction; this is possible because E has provided information of lower order, and S, if he is not already prepared to use it, will quickly learn to do so. See the later discussion of "reduction of order."

Demonstration of Fifth Order

We reverse the sequence of our presentation (definition, demonstration) at this point in order to show experimentally the need to carry our analysis beyond fourth order. In postulating fifth-order control systems, we are only saying that we think we see orderliness in the selection of fourth-order behavior patterns, and that this orderliness cannot be ascribed to the N-system or follow from our definitions of that system. Let us demonstrate fifth-order behavior, and then discuss its position in our model.

E requests S to track his finger again as before, but now E alternates between two different sequences. For example, one sequence might consist of tracing a circle clockwise, the other might consist of tracing another circle counter-clockwise. Let the two circles join to form a figure eight: one circle above the other. The upper may be designated U; the lower, L. If E produces any fixed combination of U and L (e.g., U,L,U,L,L,U,L,-U,L,L, etc.), S will eventually learn it as a single long sequence, and demonstrate fourth-order reaction time. If, however, E establishes a general relationship in his own mind which will produce an ever-changing sequence, then no fourth-order system could learn it (because it never repeats). Under these circumstances, the highest order of system which could track at all would be S_3, and S would demonstrate the jerky tracking characteristic of third order, just as though he were following a random target pattern. E can check this either by producing a random alternation of U and L sequences, or by setting up a random or very complex spatial pattern.

If E uses some fixed relationship, however, to determine whether the next sequence shall be U or L, S will eventually perceive what that relationship is, and will be able to track smoothly and change sequences at just the right time. Let us say the relationship can be perceived in this example: U,L,U,-L,L,U,L,L,L,U,L,L,L,L,.... There is no one word for this relationship among sequences, but it can be described as "increasing the number of L sequences by one after every U sequence." A person with fifth-order systems can perceive this relationship directly, whether or not he verbalizes it.

The relationship described has a reverse: one can *decrease* the number of L sequences each time instead of increasing it, provided that one starts with more than one L sequence. Thus at some point in the fifth-order tracking process, E can switch to the reverse relationship. Naturally, the first time he does this S's smooth tracking behavior will degenerate to third order or just go to pieces altogether. After some practice, however, S will have learned both relationships, and can switch from one to the other as soon as he sees that the change has occurred. Now fifth-order reaction time will be observed, and it will prove to be considerably slower than S's fourth-order reaction time. The reader may like to test our assertion that complexity is no determinant of order; try this elementary fifth-order switch: U,L,U,L,U,L,U,L,L,U,L,U,L.... Check S's reaction time both to the double L and also to a sudden stop in the complex sequence, U,L,L,U,U,L,U,L,L,U,U,L.... Remember to give sufficient practice so that S is reacting as fast as he can.

Incidentally, in the string of symbols U,L,U,L,U,L, one tends to perceive pairs U,L; if a double letter occurs, one switches to perceiving it as L,U. In reading a long alternating string, one can switch back and forth intentionally, and the effort required to do so is quite apparent: U,L,U,L,U,L,U,L,U,L,U,L,U,L,U,L,-U,L,U,L,U,L,U,L,U,L,U,L,U. This is probably as primitive a fifth-order phenomenon as is possible to perceive. Of course the two sequences, U,L, and L,U, are fourth-order perceptions as written (at the most); it is the act of changing interpretations that reveals the presence of a fifth-order system.

Fifth Order

Fifth-order systems perceive lower-order information in terms of *relationship*. This word is almost as explicit as the term "sequence," because most of its meanings actually apply at fifth order. In an arrangement (f_3) of dots, one can perceive many relationships: separation of any two dots, a triangle, relative size of the arrangement, distance (imagined) from the viewer, and so forth. In the pair of f_4s "man running" and "another man running" one can perceive "chasing," "racing," "fleeing," "greeting," and so forth.

It is important to grasp the fact that relationship and arrangement are perceived at *different orders*. At third order, every different arrangement of three dots yields a different f_3; at fifth order, a system evolved to perceive in terms of triangles might see the *same* triangle-relationship in those same sets of dots. Likewise, although one might call a sequence a "temporal relationship," we do not use this sense of the term, because a sequence-sensing system responds only to a specific range of sequences among a specific set of lower-order signals. While a sequence-signal may represent *one instance* of a temporal relationship, it does not represent the relationship itself. To a given fourth-order system, the occurrence of the double L in the above strings of symbols would be only a momentary disturbance of the sequence, and it would quickly see that the "proper" alternating sequence was still occurring, provided no control action was required. The fourth-order Feedback Function recognizes only that an L should be followed by a U, and a U should be followed by an L. It does not "group" these elements.

Fifth-order perceptions relate to areas of wide psychological interest; a man's relationships with other men can be seen as his role, his occupation, his status, and so forth; man-machine relationships can be seen (operator of a machine, victim of buzz-saw, inventor of a device), and the relationships among one's own subsystems can be described (self-respect, conflict, coordination). Interpersonal relationships and group dynamics are fifth-order subjects of study. Communication is conceived by us as essentially a fifth-order activity.

Sixth Order

Our present concept of the nature of sixth-order systems is still rather vague. As previously, we see orderliness in the choice of goals for the highest order described, fifth, and we therefore suspect the presence of higher-order systems and higher-order goals.

Our best guess to date about the nature of sixth-order perception is that f_6s represent variables pertaining to organization or orderliness, which are aspects of *systems*. Thus we say that S_6 perceive and control the nature of *systems* the elements of which are specific relationships and lower-order entities. The thing we call "personality" may be a sixth-order perception; other examples might be a symphony, a government, a self-concept, a scientific discipline, and a mathematical proof.

A "fact" may be partially defined as a perception which does not cause an error in a sixth-order system; certainly any perception which creates a sixth-order error is treated as nonfactual at first. A magical trick, a reformed criminal, electron diffraction, and a host of other phenomena have caused more than one person to doubt that he has perceived correctly. This does not mean that one doubts *having perceived* such a phenomenon; it means only that because the perception makes him see a system different from his reference-system, he looks for added information which will give a different fifth-order or lower perception, thus correcting the sixth-order error. He looks for the black thread holding up the magic wand, the secret vice which will mar the perfect behavior, the error in technique which produced the seeming diffraction rings. Sometimes the required information is found, sometimes not. The nature of one's sixth-order systems and the activity of one's N-system will then determine whether the required shift in sixth-order perception and reference-level will occur.

So far we have not found any clear demonstration of sixth-order reaction time to go into our Portable Demonstrator.

N-System

The N-system, it will be recalled (Part 1), senses the discrepancy between a set of intrinsic reference-signals and a set of perceived intrinsic signals representing critical organismic variables. Some of these variables are probably the signals associated with drives, while others may represent more subtle conditions, such as average stimulus input rate, or mean error signal in the hierarchy. We do not know how specific we have to make the reorganizing activity which is the output of the N-system. It may be a random effect randomly distributed in the hierarchy, it may be localized in regions where error-signals exist, it may operate according to some rule more efficient than a random shuffling of thresholds (as suggested by modern learning-machine experiments on computers). We are confident that we can learn more about these properties of the N-system, but we do not know much yet.

The N-system's activity has a very characteristic effect on behavior. When a reorganization is taking place, *a formerly skilled behavior deteriorates*. However, because not all systems are undergoing complete transformation all the time, behavior as a whole during reorganization still has some organization. If one is changing his concept of a skill, the overall coordination may go to pieces, but he will still show the ability to carry out specific sequences and to control configurations, and so forth. What is called "trial-and-error" behavior may often be not an organized search for a new pattern, but the automatic result of changes in high-order organization, necessarily resulting in alterations of reference-levels in lower-order systems.

In a complex-learning experiment by the authors (still in process) we have established a task in which the S must learn five orders of skill successively in order to meet the requirements outlined in the instructions. Achievement of each new order is soon marked by a plateau in the graph of reaction time against response number. We regularly observe that just before reaction time drops to a new plateau, it begins to vary and becomes longer; the graph shows a great deal of "noise" just before a drop. We take this to be evidence of N-system activity.

Subjectively, the N-system is responsible for the phenomenon called "insight," the "aha" reaction when one suddenly perceives a pattern in lower-order information which he has never seen before, or has never connected with the particular circumstances. This sort of insight is not necessarily helpful or harmful; it is merely a new organization of perception. With the ability to perceive a new pattern, one may experience extensive changes in equilibrium in many subsystems, for the better or for the worse. In the usual case, insights which are not useful are quickly discarded because they create conflict or tend to increase intrinsic errors. For example, it might occur suddenly to a golfer that he might use his putter like a billiard cue and have a better chance at a difficult putt, but a moment spent imagining doing so would reveal the fifth-order and sixth-order ("That's not golf!") conflicts which he could expect. Therefore this behavior pattern is not selected as a reference-level on the golf course.

Notice that the N-system never adds new information to the system in the sense of providing a specific answer to a problem. *It merely alters the properties of a system, thus changing the transformations applied to existing information.* It is possible for the learned hierarchy to ignore a new transformation, if higher-order systems perceive that use of it would not achieve the required higher-order perceptual fields. Furthermore, a new perceptual transformation may be such that it is of no use in present circumstances, but may be useful later, so that it appears to "lie dormant" for a time. In solving a mathematical problem, it is common to perceive the final steps which will lead to the required solution long before one has found out how to lead up to them.

Before we leave the subject of N-system, we wish to propose a definition. We have some fairly good reasons for this proposition, but for now we prefer merely to state it and explain what we mean: this is a definition of consciousness.

Consciousness, we propose, is the state of the feedback-function in a subsystem in the hierarchy which is being affected by the output of the N-system. Thus the same subsystem can perceive and control a variable either consciously or unconsciously, depending on whether the N-system is actively connected to it. The objects of consciousness are interpreted by the feedback-function of the conscious learned system; the subjective experience is that of seeing these interpretations *in the objective perceptual field*. A conscious S_4 perceives that sequences are going on in its environment. A conscious third-order system sees an environment composed of arrangements. A conscious fifth-order system sees a set of relationships.

Consciousness is not differentiated from order to order. One can be conscious simultaneously of a number of different orders of perception. Our language reflects this property by compressing several orders of percept into single sentences: "The little square is inside the big square" stands for perception of several relationships (little, big, inside) and several configurations (the squares). Furthermore, the sequence in which the words are placed identifies how the elements are to be arranged in the "inside" relationship.

The properties of the N-system give consciousness some interesting properties, as we define it. The reader may find it intriguing to consider the effect on a skill normally carried out unconsciously when the system that controls it goes into the state we call consciousness. The reader may also wish to give a detailed verbal description of how he ties his shoelaces while he is doing it.

Reduction of Order

We have invariably found that human Ss and perhaps animals as well will attempt to use the lowest order of information available that will suffice for doing a task. In design of a sequential task for measuring S_4 properties, one must be careful that there is no element in the sequence which by itself provides enough information for successful completion of the task. We have asked Ss to make differentiating responses to two sequences of spot deflection, "left, left," and "right, left." Nearly all Ss showed third-order reaction times (0.3 sec., approximately) instead of fourth (0.4-0.45 sec., approximately), because they learned to see the initial spot as an element in the total arrangement of cues on an oscilloscope. If the initial jump filled the blank space to the left, they gave one response, and if it filled the space to the right, they gave the other. No attention had to be paid to sequence at all. It is very difficult to avoid giving such lower-order information. This order-reduction effect may account for what is termed "stereotypy" in learning situations, where an S will continue to give a response even though changed circumstances make it inappropriate. Since S is attending to the problem at a lower order than E intended, he fails to notice that the higher-order situation has changed.

Order reduction is carried out by human beings in another interesting way, through use of symbols. The reader will remember the lower circle, L, and the upper circle, U, employed in the fifth-order Portable Demonstrator. We represented these two sequences by letters, and then proceeded, later on, to use these letters simply as third-order objects. The letters were actually order-reduced representations of sequences, but could be used any other way we pleased. What makes the difference is the set of rules one uses for manipulating the third-order objects. If they are treated as algebraic variables, then they are manipulated according to the (fourth-order) rules of algebra. If they stand for sequences, then they are manipulated according to a different set of rules. This procedure is very common in mathematics; letters often stand for *operators* which are actually sequences of manipulations, and there is a set of rules for the algebraic manipulation of operator-symbols, different (but nevertheless still fourth-order) from the ordinary rules of algebra.

By such use of symbols as order-reduced representations of higher-order perceptions, one can build verbal or logical structures with many more levels than there are orders of perceptions. Of course the rules relating one level to another will still be of six or fewer types, corresponding to the transformations among human orders of perception. By such order-reducing techniques, human beings can construct symbolic variables representing combinations of events in the second-order perceptual field which they cannot perceive directly —they cannot build feedback functions of sufficient complexity or sufficient accuracy to respond directly to these combinations as abstract variables. We have employed this technique constantly in building our behavioral model. We cannot, however, take credit for discovering the technique—most language is an order-reduced representation of experience.

Conflict Theory

Consider two systems of order n, both controlling the reference-level of a system of order $n-1$. Generally, each nth-order system will also control other sets of systems of order $n-1$. Often, the nth-order systems can achieve their respective reference-levels independently of one another, even though they share some systems at order $n-1$, because other systems can be adjusted to compensate for potential conflicts. But in the case where the two reference-levels at order n demand mutually contradictory settings of r at r_{n-1}, and the common subsystem is essential to both higher-order systems, conflict occurs. Likewise, two higher-order systems can often be simultaneously satisfied by finding a suitable lower-order system; one can satisfy the desire to ride a bicycle and the desire to go downtown by employing the same skill. But, if no common system with this property exists, then both higher-order systems remain unsatisfied, and any attempt to set appropriate reference-levels at lower order will result in conflict.

Conflicted FBCS are in a condition in which correcting the error in one system increases the error, and hence the corrective efforts, of the other system. If both systems were good control systems in the first place, they will react strongly to even moderate errors, and hence when in conflict will tend to send out extreme output signals. This does not necessarily mean *energetic* outputs, but only that if the opposition were suddenly removed, behavior would follow some extreme pattern.

The common subsystem will behave as though its reference-level were set at a compromise value which we term the "virtual reference-level," and will act like any other control system. But because the controlling systems are near or at their limits of output, the controlled system will appear to have a fixed reference-level which does not change with circumstances. For all practical purposes, the two controlling systems have been removed from the organism's set of environment-controlling systems, and are serving only to generate a constant reference-signal in the controlled system.

If conflict is severe enough to drive the conflicted systems to their limits of operation, another effect may occur. Feedback systems lose their resistance to disturbance when driven to their limits, but they also enter a very non-linear region of operation, in which all their important characteristics change. A common result is instability, which shows up as oscillation. Thus the system might oscillate, and behavior would show what is called "vacillation."

A pseudo-threshold effect can be seen when a system limits; the error it has been holding near zero suddenly begins to increase when the system reaches maximum output. If an older system exists, which operates to keep a similar error-signal near zero but which is much cruder, requiring a greater error to produce output, the older system will come into action as the error increases, and behavior will be typical of the older system. This is in part our explanation of "regression" and its connection with conflict and extreme stress.

Conflicts can be removed by altering the properties of one or the other controlling system, by altering the reference-levels of either system, by switching one system entirely to the imagination-connection [the mechanism (See Part I) of fantasy, wish-fulfillment, and closure], by introducing a third system to conflict with the unwanted system (suppression), and a few others. In general the removal of the crippling effects of conflict by means *other than* a change in reference-level by a still higher-order system or a realistic change in properties via N-system action, is the mechanism of what is known as the "defense of the ego." All of the classical psychoanalytic defenses can be seen easily as solutions to the general conflict situation outlined above.

It should be remembered that conflict only removes the higher-order systems from action: the commonly-controlled lower-order system remains in action, and actively maintains behavior at the virtual reference-level created by the different control-signals. This is one form of "resistance" and is what makes it seem that the system is resisting all change, at least in the conflicted area.

If one arbitrarily forces behavior toward one or the other of the contradictory higher-order goals, the output of the corresponding system will decrease toward neutral as its error-signal decreases. But the other system will still be producing maximum output, therefore it will appear that S has suddenly begun to take higher-order action against the disturbance. If the conflict is mild, so that neither system has quite reached maximum output, this effect will be more pronounced. An external agency forcing behavior in the "right" direction will find that S's own motivation in that direction relaxes and his efforts in the opposite direction increase, just as though he were seeking to maintain exactly his present state. This is why the "will power" and "authoritarian" approaches seldom have any lasting effects, and why it appears often that people are actively keeping themselves in unpleasant conditions.

In general, it is plain that the system which is actively maintaining behavior constant is one order lower than the conflicted systems. This is a useful rule to keep in mind if one is a therapist. A person who is in the grip of a compulsive sequential behavior-pattern at fourth order is conflicted at fifth order, not fourth—a fourth-order conflict results in actively maintained stasis, not action. Likewise, a person who shows rigid behavior toward other people is actively and efficiently maintaining his fifth-order interpersonal relationships at a frozen reference-level, and the conflict is at sixth order, not fifth. He has problems concerning his concept of systems; his own, society's, or other people's. It does no good to alter a person's behavior at an order lower than the level of origin of the conflict, except for purposes of safety or survival. The conflict remains and will be expressed differently at the lower order. The paralyzed leg turns into a paralyzed arm; the hatred for father becomes a hatred of money; the compulsive handwashing becomes compulsive bead-telling. Only a change in the systems which are fighting each other through the lower-order systems will have a permanent effect.

Finally, by our postulated definition of consciousness, putting both or all the conflicted systems into the state of consciousness is the only way to start the N-system to work changing them.

Demonstration of Conflict

The Portable Demonstrator can readily exhibit some rather striking examples of conflict, along with the various possible results. If E joins his hands, aligning the two forefingers to provide a single indicator to be tracked, S can readily acquire the fourth-order system needed to track some repetitive movement with his single forefinger. After continuing until S has clearly established his fourth-order system, E *separates* his two hands, moving each in a different manner. Thus E now provides *two*, incompatible, fourth-order reference-signals. The movement should be simple, such as a simple circular sequence, or moving the two fingers in opposite vertical or horizontal directions, and S can be practiced in both sequences. Indeed, as with all the Portable Demonstrator presentations, it works at least as well when S knows just what to expect. In this last demonstration, however, we cannot predict *which* of the various possible responses to the conflict situation will be selected. Commonly S will demonstrate the virtual reference-level, especially if the conflicting signals are equal and opposite. However, it may be possible for S to select one of the two and ignore the other—this is difficult. There are many variations of this demonstration, and the analysis of several of them should be quite instructive for the reader.

A Statement of Values

Perhaps it is fitting to close this section of our paper by a brief statement of how we view the properly-operating FBCS hierarchy.

In the optimum system, no significant conflict exists, so that all systems important to behavior are free to operate over their full range without internal opposition. Likewise, the concept which the system has of itself must include knowledge of the properties of the N-system and the signs of its action, so that the N-system remains free to keep intrinsic errors minimized, and so that the results of N-system action can institute change anywhere in the hierarchy without undue self-preservative action on the part of existing systems. Thus it is capable of modifying its systems as rapidly as changes in its environment may require.

If the organism is in this state, it is performing properly; there is nothing wrong with it. The person perfectly organized in this respect can still fall into conflict with himself, but the N-system is capable of finding solutions if they exist. The person is still subject to the limitations of his environment, to the distortions of false information, and the illusions inherent in the geometry of perception. The person may be a saint or a sinner, but he will not be mentally incapacitated.

There is no morality inherent in our theoretical structure, although the phenomenon of moralizing can be easily described in its framework. The definition of an optimum FBCS hierarchy reflects our personal preferences—we prefer to see people performing "up to specs," regardless of what they choose to do, and it is toward this end that we choose to work. We also choose to try to persuade other people to accept this goal to see how it works out. Our value choice implies many detailed attitudes toward human personality and interaction. The optimum system, for example, can be controlled (in a basic sense) only from within; its ultimate determinant of action is satisfaction of its intrinsic reference-levels, whatever they may be. To the extent that a person can be controlled by outside agencies, to the frustration of what he originally wanted, he has something wrong with his internal organization. Of course a good deal of what is termed "control" of behavior is not control at all, but the normal action of independent systems in the process of satisfying their own intrinsic states or learned goals. With respect to the natural organism, the only way to control it is to get control of the means for satisfying its intrinsic state. (This is exactly what is done in much animal experimentation.) This is very dangerous when tried on human beings, because the putative controller is likely to find himself being treated as a disturbing variable in process of being removed.

We believe that all human behavior is essentially based on the individual's experimentation. He would have to try moral values, facts, methods, and so forth to test them for their effectiveness in the ultimate task, that of maintaining his intrinsic state. He would even have to learn, if he could, just what constitutes his own intrinsic state, because this information is not built into his learned systems. His naturally acquired concepts of the details of human behavior are the *content* of his hierarchical system and are all learned. His concepts of social interaction, personality, and all the rest of his attitudes and behaviors were invented by someone, and reinvented, and taught. If the human lot is to be improved, a better picture of "the good" and of "right and wrong" must be invented, and people must also explicitly recognize these ideas as inventions up for critical test. A person must beware of setting up any ideal in his mind as sacred and untouchable, for by so doing, he sets up an automatic mechanism to counter the effect of his *N*-system, and guarantees a halt in his development.

Summary

A two-part theory of human behavior has been presented; Part I deals with a general conceptual model based on a hierarchical arrangement of negative feedback control systems, of the type in which the control loop includes portions of the environment. Each level of system controls the level below by specifying reference-levels for the controlled systems. Part II outlines applications of the feedback principles to behavior, and introduces six hypothesized levels of perceptual variables associated with human feedback control systems. These levels range from spinal reflexes (First-Order Systems) to systems which perceive and maintain orderliness and system concepts (Sixth-Order). An organizing system is described.

[1971]

A Feedback Model for Behavior: Application to a Rat Experiment

Stimulus-response laws can be rendered trivial when environmental feedback exists from R to S. An input quantity (q_i), defined as the actual environmental quantity or event that leads to a response, is a function of both the applied stimulus (S) and the feedback from the related ongoing behavior (R): $q_i = h(R, S)$. The observed behavior is dependent on actual input stimulation via the organism function: $R = g(q_i)$. Hence $R = g(h(R, S))$, and not $R = g(S)$.

Analysis of a shock-avoidance experiment done by Verhave illustrates a method for taking environmental feedback effects into account; the resulting model fitted to the behavior of one rat predicts the behavior of another rat in an altered experiment with an *RMS* error of less than one barpress per minute. Graphical solutions to a range of possible functions g and h (as above) show why this type of experiment reveals more about the experimental apparatus than about the rats.

When environmental feedback is significant (and negative) one must characterize the organism's actions as behavioral control of stimulation and not stimulus control of behavior.

It has been recognized that the output of an organism, the muscle forces and their consequences which appear as an organism behaves, must influence the inputs, the sensory

Copyright 1971 by the International Society for Systems Sciences. Reprinted with permission from *Behavioral Science* 16(6), November 1971, 558-563.

events which appear to give rise to specific actions. The traditional view of behavioral organization, however, provides no way to analyze a situation in which this effect of output on input occurs essentially simultaneously with variations in the input: a situation in which the stimulus and the response can be characterized only over rather long periods of time and hence necessarily overlap. If the feedback from output to input is immediate and strong, then the applied stimulus or manipulated variable coexists in part with the effects of the response, and the effective input influencing behavior must be a joint function of both.

When output affects input in this way, feedback exists. Attempts have been made to incorporate this sort of feedback into the traditional stimulus-response concept of behavior, but in all such cases of which the author is aware the feedback effects have been treated as if they could be separated from the effects of applied stimuli, for example, by alternation in time. Indeed, only by doing so is it possible to preserve the concept that behavior is strictly a function of input stimuli. As soon as one admits that the inputs are at least in part determined by the ongoing outputs, a different picture begins to emerge, that of an organism which actively controls the status of its input variables. The present paper is concerned with a very small aspect of this new view of behavioral organization.

If a response affects an organism at the same time that a stimulus affects it, then one can no longer say that the effective stimulus is the same as the applied stimulus. Where the traditional model is given as S-O-R, indicating that the stimulus produces a response via mediating functions of the organism, we must substitute a model that provides a way for both applied stimulus and response to affect some input quantity q_i; the diagram which indicates this feedback relationship must contain a closed loop, as follows:

$$S \longrightarrow Q_I \longrightarrow O \longrightarrow R$$

The input quantity q_i is defined so that the response measure is some function of that variable; q_i, however, is a function of two variables: the response and the applied stimulus, or independent variable. If we let g represent the organism function which makes R depend on q_i, and h the functional dependence of q_i on both R and S, we have two functions, the first, g, associated strictly with the organism's inner organization, and the second, h, associated only with environmental laws:

$$R = g(q_i), \text{ and}$$
$$q_i = h(R, S)$$

Obviously the response will still show a relationship to stimuli, but now the presence of feedback causes the observed dependence of response on stimulus to be misleading;

$$R = g(h(R, S)), \text{ not } R = g(S)$$

In order to extract from an empirically observed stimulus-response law the function which pertains to the organism, it would be necessary to solve the equation for R. As the equation stands, the observed relationship describes not just the organism, but the organism plus some properties of its surroundings through which the output has effects on the input. In short, all experiments in which a response can have effects concurrent with the effects of an applied stimulus give a more or less spurious picture of the actual organism function that is involved. This would apply in particular to all operant conditioning experiments. Feedback renders a simple input-output analysis irrelevant.

To illustrate the preceding concepts, I will analyze an experiment done by Verhave (1959). In this experiment, rats were required to press a lever a fixed number of times within a specified interval. If the required number of presses was executed in time, the interval timer would automatically reset starting a new interval. If less than the required number of presses occurred before the end of the predetermined interval, a shock would be administered, and the timer would again reset to start a new interval.

After sufficient practice for a given setting of the interval timer and a given number requirement (constant during one experiment), rats would approach some equilibrium rate of pressing. Thus a relationship was explored with the setting of

50 Living Control Systems

the interval timer as the independent variable and the equilibrium rate of pressing as the dependent variable. Each experimental point was the average of three different four-hour averages of rate of bar-pressing. Scatter among the three determinations for a single point was on the order of one press per minute.

As one might expect, when N presses had to be executed within 15 seconds (the shortest interval used), the average rate of pressing was much faster than it was when the same number of presses had to occur within 5 minutes (the longest interval). The final relationship is a fairly regular curve that is amenable to mathematical approximation. The observed relationship is indicated in the first two columns of Table 1, for a number requirement of N presses in I minutes, with $N = 8$.

The average rate of bar-pressing was always much faster than the rate actually required in order to avoid shock. The reason for this can be seen in the variations in bar-pressing rate; even with the average rate at a value fast enough to avoid shock, on some trials the random variations in rate were sufficient to delay the 8th press enough after reset of the timer to permit a shock.

Table 1. Theoretical Values Fit to Data

Animal	Rat 17			Rat 20	
Number requirement, $N =$	8			1	
Hypothesis, $q_i =$	p_s		$r_s = (p_s/I)$	p_s	
I, min.	r_p, obs min^{-1}	r_p, calc	r_p, calc	r_p, obs	r_p, calc
1/4	70.8	69.66	73.85	13.29	12.32
1/3	53.6	55.57	55.39	10.42	9.59
5/6	26.8	28.15	22.15	5.93	6.82
1-2/3	14.3	14.42	11.08	3.72	4.20
2-1/2	11.2	10.12	7.42	—	—
5	5.4	5.48	3.68	—	—
RMS Error, min^{-1}		1.00	3.24		0.82

Column 4: $r_p = (550/I)\text{ierf}[1.40(r_p I/8 - 1)]$
Column 3: $r_p = (546)\text{ierf}[0.965(r_p I/8 - 1)]$
Column 6: $r_p = (546)\text{ierf}[0.965(r_p I/1 - 1)]$

Constructing the Feedback Analysis

The random variations in bar-pressing rate (random with respect to the subsystem involved in this behavior) provide a way to express the effect of output (rate of pressing) on input as a smooth function. Figures 1 and 2 indicate how this can be done.

In Fig. 1, a curve is shown representing the probability density function for the interval occupied by N presses, N/r_p. A vertical line indicates the setting (I) of the interval timer. The shaded tail of the distribution curve to the right of the vertical line has an area proportional to the probability of a shock for an interval I. For high rates of pressing the peak of the curve will move to the left, lowering the probability of getting a shock; for lower rates of pressing (longer interval occupied by N presses) the probability will increase.

Thus if the shape of the distribution function is known, the effects of both the response (the rate of pressing) and the

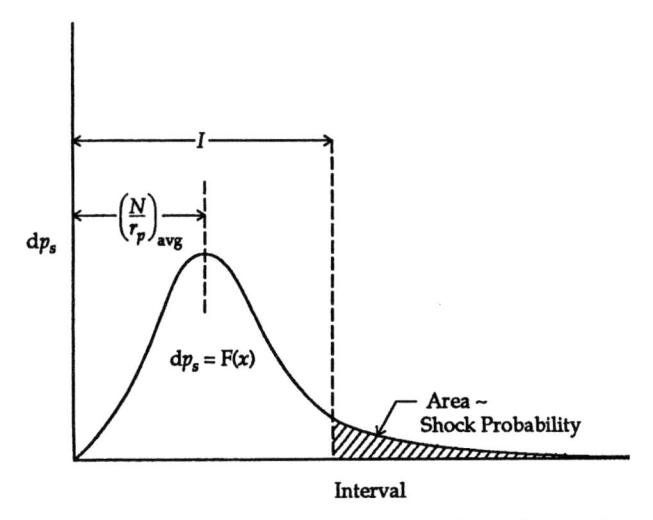

Fig. 1. Effect of bar-pressing rate on shock probability. Groups of N presses occurring at an average rate r_p occupy an average interval of N/r_p minutes, with some variability in rate of pressing. Those groups of N presses which take longer than I minutes result in a shock. The probability of a shock per trial is thus proportional to the area of the distribution curve to the right of the interval I.

stimulus (the independent variable I) on the probability of a shock can be calculated. In this experiment the actual temporal distribution of rate of pressing was not recorded, so it will be necessary to assume a form for this curve.

An obvious factor in determining how much effect a change in response has on the input quantity is the width of the distribution curve. For very narrow widths (regular, rhythmic pressing), a slight change in rate of pressing can have a drastic effect on the rate at which shocks occur. This corresponds to a strong feedback connection, and even if the feedback is negative (it is), too high a sensitivity at this point can lead to instability and oscillation—which is, in fact, sometimes observed in similar situations as bursts of bar-pressing separated by intervals of no pressing.

The analytical form chosen to represent the distribution function should therefore contain at least one parameter representing the width of the distribution. The normal distribution is one such form, and will suffice for this demonstration. (The Poisson distribution is often used in this sort of situation, but the standard deviation of that function is simply equal to the mean and so is not available as a parameter to be determined from the data. In addition, it provides a very poor final model; the tail of the curve falls off far too rapidly.)

Since the probability of occurrance of a shock is an area in Fig. 1, it can be plotted by integrating the curve. The equation expressing this probability under the assumption of a normal distribution is

p_s = ierf $A((X - N/r_p^*)/(N/r_p^*))\,|_{x=I}^{x=\infty}$ where
p_s = probability of a shock in time I
N = number of presses required within I minutes
r_p^* = mean rate (average value) of pressing, presses/minute
A = width parameter, to be determined (inverse standard deviation)
ierf = integral of the error function of...

The term N/r_p^* is the average interval occupied by N presses at r_p^* presses per minute. The argument of the integral error function is the fractional difference between this interval and the critical interval I. The integral error function must be evaluated from minus infinity to I which is nonsensical, but the left half of this distribution does not have any effect since in fact $N/r_p{}^*$ is always much less than I; we can simply assign a probability of 0.5 for the integral from minus infinity to $N/r_p{}^*$. As it turns out, the normal distribution seems to have about the right shape where the experimental data lie. No doubt, the actual distribution would be somewhat different.

Now consider Fig. 2. Illustrated (not plotted) is the dependence of probability of getting a shock on the rate of bar-pressing (rather than on $N/r_p{}^*$). Also shown is a hypothetical rat model, $g(p_s)$. In the absence of better information, we will represent the function g by a straight line approximation. The slope of this line (with the horizontal axis as the dependent variable) characterizes the average sensitivity of the animal to shock probability, or to any variable depending on that probability, in terms of presses per minute per unit probability. This amounts to hypothesizing the probability of a shock in

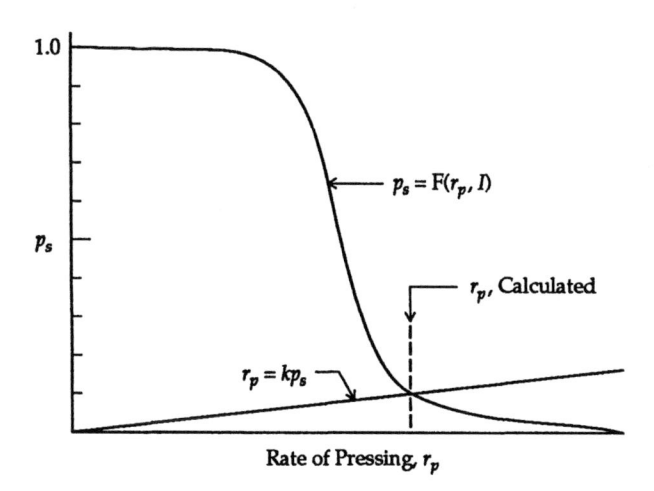

Fig. 2. Simultaneous solution of system equations. From Fig. 1 can be computed the dependence of the probability of a shock (p_s) on the rate of pressing (r_p) and the critical interval (I), as shown by the upper curve. The lower (straight) line is the assumed rat function: the rate of pressing is assumed proportional to the probability of a shock. Note that for the rat function, the dependent variable is on the horizontal axis. At the intersection is found the rate of pressing that satisfies both relationships.

the interval I as being the actual input quantity, q_i, to which the animal is responding, and approximating the nonlinear rat function as a simple constant of proportionality linking q_i to the rate of pressing. This constant of proportionality is the second adjustable parameter in the whole model, which is now given by two equations:

$$r_p{}^* = kp_s$$
$$p_s = \text{ierf}\,((A)(I - N/r_p{}^*)/(N/r_p{}^*))$$

Fig. 2 was plotted to show the graphical solution to these equations. The solution is the intersection of the straight line and the curve.

The analytical (that is numerical!) solution is given in Table 1, Column 3, with the parameters A and k selected for best fit. Clearly, no serious errors have been made since the model curve follows the data within about one press per minute. Column 4, incidentally, illustrates the fit when q_i is hypothesized to be r_s, the rate at which shocks occur, or p_s/I. The fit is enough worse to cause this hypothesis to be ranked lower than $q_i = p_s$.

There seems to be a great deal of luck involved so far, considering the loose nature of the assumptions. To stretch this luck even further, let us see what happens when the equation is applied directly to another experiment, involving a second rat and also involving a change in the number requirement from 8 presses in the interval I to 1 press. The constants A and k are left as determined in the first experiment. Columns 4, 5, and 6 of Table 1 show the experimental data and the prediction generated from the equation. The only change in the equation was to change N from 8 to 1, yet the model fits the new data within 1 press per minute, again within the scatter of repeated 4-hour determinations for each point.

Discussion

This rather surprising result is not, as may appear, the product of a long search for analytical forms that would give an acceptable fit. In fact, there are many forms similar to those chosen which serve nearly as well. The actual assumptions involved are few indeed: the shape for the distribution curve, and the linear proportionality assumed for the rat function. Why should the model derived in so simple a manner represent so well what must be a very complicated situation? An answer to that question can be seen by plotting more curves and seeing how changes in hypothesis ought to affect the model. In Fig. 3 are plotted one hypothetical distribution integral and several rat functions having widely differing slopes. It is immediately obvious that for a given setting of the interval timer, roughly determining the 50 percent probability point, the range of possible solutions is restricted to a relatively narrow range of rates of pressing. The actual rate of pressing observed will not be affected nearly as much as one would expect from knowing, for instance, that one rat is 20 times as sensitive to shock as another.

In Fig. 4, a single rat function is plotted, and several distribution functions of varying width are shown. Here it can be seen that for very narrow distributions, the slope of the

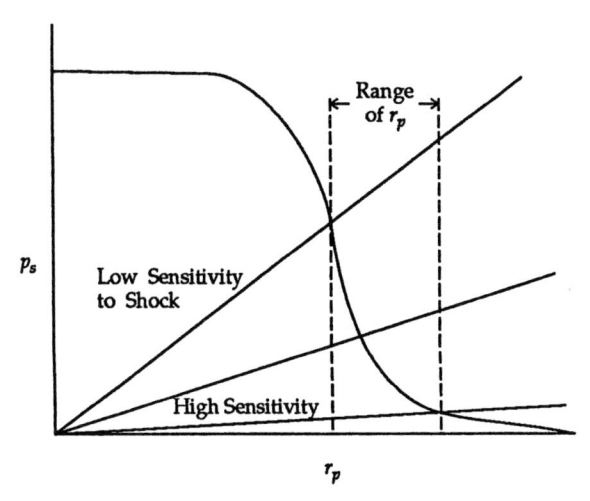

Fig. 3. Effect of variations in the rat's sensitivity to shocks. A very large change in the rate of pressing produced by a given shock probability (a 1400 percent change) results in a far smaller change in actual rate of pressing observed (37 percent).

rat function will have scarcely any effect on observed rate of pressing; the width of the distribution curve is a principal determinant of the final observed relationship.

Finally, in Fig. 5 are shown distribution functions of about the same width for several different values of the critical interval I. Also shown are several possible rat functions, ranging from a linear function to a wildly nonlinear function that reverses slope twice. The rate of bar-pressing is obviously affected relatively little either by the average sensitivity of the rat to shock or by the detailed nonlinearities of the rat function.

From these considerations several generalizations can be drawn. First, in bar-pressing situations such as this, one must expect that the lower the noise (random variations in rate), the less the experiment is going to reveal about the actual input-output function of the organism itself and the more the result will be dictated by the experimental setup. The more precise the data the less it will mean! Second, for relatively narrow distributions, the experimental data will not reveal very clearly (if at all) the actual form of the organism's re-

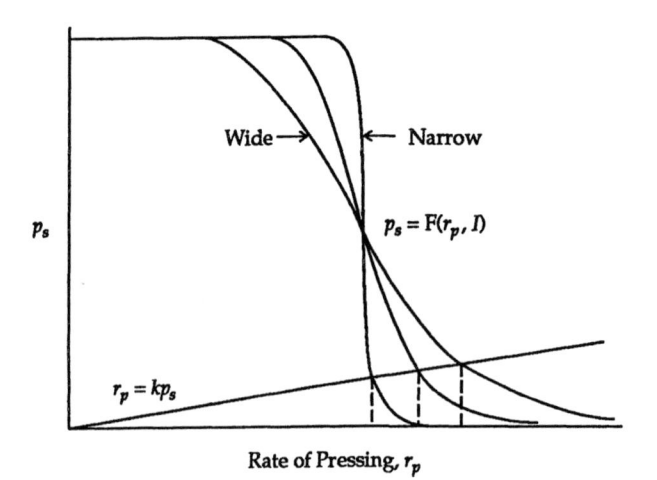

Fig. 4. Effect of variability in rate of pressing. For very narrow distributions (rhythmic pressing) the rate of pressing is nearly independent of the rat's sensitivity to shock (k). Wider distributions permit somewhat larger effects from variations in k.

sponse to the variable causing its behavior. Going to the other extreme, if the distribution function is very broad, the feedback effects will be weak—but by the same token, the data will be extremely noisy, and once again the characterization of the organism itself becomes difficult, or impossible. Only for some intermediate region of randomness in responding will it be possible to obtain even poor information about the organism that is supposedly being studied. The presence of feedback makes this sort of experimentation simply the wrong approach.

None of this would be a surprise to a servomechanisms engineer. Any time there is negative feedback from output to input of an active and reasonably sensitive system, he would expect the observed input-output relationship to be determined primarily by the effects of the feedback path which in this case is characterized by response variability and the experimental apparatus. He would see the input quantity q_i as being under control not of the stimulus but of the system, the organism; the stimulus clearly appears in the role of a dis-

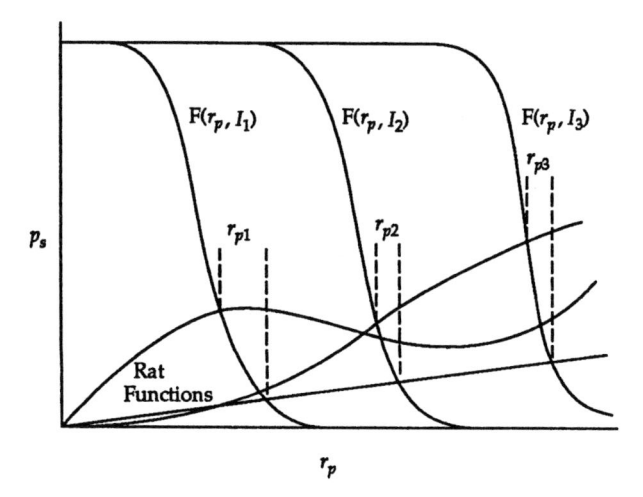

Fig. 5. Relative effect of I and Rat function on rate of pressing. The setting of the interval timer, for a given distribution, has far more effect on r_p than does the form of the rat function. Thus the experiment reveals more about the experimental apparatus than about the rat.

turbance, the effects of which will be nearly cancelled by the feedback effects. In fact, the rate of shocking is very low in this sort of experiment. Compared with the rate that would occur if there were no responses, it is practically zero. The control is excellent, and one can even assert that the set point of this servomechanism is zero. Changes in the independent variable I have little effect on changing the probability of a shock, away from the set point.

While the stimulus-response concept of behavior in this situation is misleading, experiments of this sort can be used to extract interesting information about the organism. The most important quantity that must be understood in this kind of feedback analysis is q_i, the input quantity to which the organism actually responds. In the present analysis a clear difference is seen between the fits of the curves when two different hypotheses for this quantity are chosen (p_s vs. r_s). It is reasonable to suppose that the hypothesis which gives the better fit is the closer to the actual nature of q_i. The present analysis suffers from the defect that the distribution curve was assumed rather than measured. If an experiment were set up to record this distribution, then it would be possible to arrive at a better definition of q_i.

Note that the nature of q_i is never self-evident. In many stimulus-response experiments a stimulus is applied and laws are stated in terms of effects of this stimulus on responses. Clearly, the ordinary stimulus used in such experiments is not the same as the quantities to which the organism is actually sensitive, and there is always the likelihood that feedback is present making the actual q_i depend not only on the stimulus but on the response as well. When feedback exists and is fairly strong, it is improper to think of the stimulus as causing the response in a simple way. Rather, one must consider that the stimulus constitutes a disturbance of the input quantity, and the response is such as to prevent those disturbances from significantly changing q_i. The preferred value for q_i is determined in the organism, not in the environment. In the experiment just analyzed, the preferred level for shock is obviously zero, and not because anything outside the rat says that shock should not be experienced.

Finally, it must be remembered that the sensory apparatus

of organisms contain interpretive apparatus: the input quantity may, in fact, be a function of many sensory inputs, and may come into existence only after several stages of perceptual data processing. Even when that is the case, a feedback analysis along the lines suggested here can enable the experimenter to arrive at a reasonable approximation of the actual aspect of the environment that the organism is regulating, even when that aspect is an abstraction like density, or relative size, or a probability.

References

Powers, W.T., Clark, R.K., & McFarland, R.L. A general feedback theory of human behavior, *Percept. Mot. Skills*, Part I, 1960, 11, 71-88; Part II, 1960, 11, 309-323. Reprinted in *General Systems*, 1960, 5, 66-83.

Verhave, T. Technique for differential reinforcement of rate of avoidance responding, *Science*, 1959, 129, 959-960.

[1973]
Feedback: Beyond Behaviorism

Stimulus-response laws
are wholly predictable
within a control-system model
of behavioral organization.

The basis of scientific psychology is a cause-effect model in which stimuli act on organisms to produce responses. It hardly seems possible that such a simple and venerable model could be in error, but I believe it is. Feedback theory shows in what way the model fails, and what must be done to correct our concepts of organized behavior.

Responses are dependent on present and past stimuli in a way determined by the current organization of the nervous system; that much is too well documented to deny. But it is equally true that stimuli depend on responses according to the current organization of the environment and the body in which the nervous system resides. That fact has been left out of behavioristic analyses of human and animal behavior, largely because most psychologists (especially the most influential early psychologists) have lacked the tool of feedback theory.

Norbert Wiener and later cyberneticists notwithstanding, the full import of feedback in behavioral organization has yet to be realized. The influence of behaviorism, now some 60 years old, is pervasive and subtle. Shaking ourselves free of that viewpoint requires more than learning the terms asso-

Copyright 1973 by the AAAS. Reprinted with permission of the publisher from *Science* 179(4071), January 26, 1973, 351-356.

ciated with feedback theory; it requires seeing and deeply appreciating the vast difference between an open-loop system and a closed-loop system.

Traditional psychology employs the open-loop concept of cause and effect in behavior; the effect (behavior) depends on the cause (stimuli) but not vice versa. The closed-loop concept treats behavior as one of the causes of that same behavior, so that cause and effect can be traced all the way around a closed loop (1). When any phenomenon in this closed loop (such as the force generated by a muscle) persists in time, effectively averaging the antecedent causes over some period, the character of the system-environment relationship changes completely—cause and effect lose their distinctness and one must treat the closed loop as a whole rather than sequentially. That is where feedback theory enters the picture. Feedback theory provides the method for obtaining a correct intuitive grasp of this closed-loop situation in the many situations where the old open-loop analysis leads intuition astray.

In this article I intend to show as clearly as I can how a new theoretical approach to behavior can be developed simply by paying attention to feedback effects. There is nothing subtle about these effects; they are hidden only if they are taken for granted. All behavior involves strong feedback effects, whether one is considering spinal reflexes or self-actualization. Feedback is such an all-pervasive and fundamental aspect of behavior that it is as invisible as the air we breathe. Quite literally it is behavior—we know nothing of our own behavior but the feedback effects of our own outputs. To behave is to control perception.

I will not try here to develop all these concepts fully; that is being done elsewhere (2). I will provide only some essential groundwork by discussing the development of a hierarchial control-system model of behavioral organization beginning with the same sort of elementary observations that led to behaviorism. I hope it will thus become evident that a fully developed feedback model can do what no behavioristic model has been able to do: it can restore purposes and goals to our concept of human behavior, in a way that does not violate direct experience or scientific methods. The human brain is

Act versus Result

Behaviorists speak of organisms "emitting" behavior under stimulus control, this control being established by use of reinforcing stimuli. The effectiveness of reinforcers cannot be denied, but behavior itself has not been thoroughly analyzed by behaviorists. Behaviorists have not distinguished between means and ends—acts and results (3)—because they have not used the model that is appropriate to behavior.

When a pigeon is trained to walk in a figure-eight pattern, there are at least two levels at which the behavior must be viewed. The first, which is the one to which the behaviorist attends, is that of the pattern which results from the pigeon's walking movements. The other consists of those movements themselves (4).

The figure eight is created by the walking movements: the act of walking produces the result of a figure-eight pattern in the observer's perceptions. The observer sees a consistent behavior that remains the same from trial to trial. He generally fails to notice, however, that this constant result is brought about by a constantly changing set of walking movements. Clearly, the figure-eight pattern is not simply "emitted."

As the pigeon traces out the figure eight over and over, its feet are placed differently on each repeat of the same point in the pattern. If the cage is tipped, the movements become still more changed, yet the pattern which results remains the same. Variable acts produce a constant result. In this case the variations may not be striking, but they exist.

As behaviors become more complex the decoupling of act and result becomes even more marked. A rat trained to press a lever when a stimulus light appears will accomplish that result with a good reliability, yet each onset of the stimulus light produces a different act. If the rat is left of the lever it moves right; if right it moves left. If the paw is beside the lever the paw is lifted; if the paw is on the lever it is pressed down. These different, even opposite, acts follow the same stimulus event.

The more closely the rat's acts are examined, the more variability is seen. Yet in every case the variations in the acts have a common effect: they lead toward the final result that repeats every time. In fact, if precisely those variations did not occur, the final result would not be the same every time. Somehow the different effects apparently caused by the stimulus light are exactly those required to compensate for differences in initial conditions on each trial. This situation was clearly recognized by the noted philosopher of behaviorism, Egon Brunswik (5).

The accepted explanation for this phenomenon of compensation is that the changed initial conditions provide "cues," changes in the general background stimuli, which somehow modify the effect of the main stimulus in the right way. There are three main problems created by this explanation. First, these hypothetical "cues" must act with quantitative accuracy on the nervous system employing muscles which, because they are subject to fatigue, give anything but a quantitative response to nerve impulses. Second, these "cues" are hypothetical. They are never experimentally elucidated in toto, and there are many cases in which one cannot see how any cue but the behavioral result itself could be sensed. Third, the compensation explanation cannot deal with successful accomplishment of the behavioral result in a novel situation, where presumably there has been no opportunity for new "cues" to attain control of responses.

The central fact that needs explanation is the mysterious fashion in which actions vary in just the way needed to keep the behavioral result constant. The "cue" hypothesis comes after the fact and overlooks too many practical difficulties to be accepted with any comfort. Yet what is the alternative? It is to conclude that acts vary in order to create a constant behavioral result. That implies purpose: the purpose of acts is to produce the result that is in fact observed. This is the alternative which I recommend accepting.

Feedback Control

Behaviorists have rejected purposes or goals in behavior because it has seemed that goals are neither observable nor essential. I will show that they are both. There can be no rational explanation of behavior that overlooks the overriding influence of an organism's present structure of goals (whatever its origins), and there can be no non-trivial description of responses to stimuli that leaves out purposes. When purposes are properly understood in terms of feedback phenomena, acts and results are seen to be lawfully related in a simple and direct way. We will see this relationship using a simple canonical model of a feedback control system.

Engineers use negative feedback control systems to hold some physical quantity in a predetermined state, in an environment containing sources of disturbance that tend to change the quantity when it is uncontrolled. Every control system of this kind must have certain major features. It must sense the controlled quantity in each dimension in which the quantity is to be controlled (*Sensor function* in Fig. 1); this implies the presence of an inner representation of the quantity in the form of a signal or set of signals. It must contain or be given something equivalent to a reference signal (or multiple reference signals) which specifies the "desired" state of the controlled quantity. The sensor signal and the reference signal must be compared, and the resulting error signal must actuate the system's output effectors or outputs. And finally, the system's outputs must be able to affect the controlled quantity in each dimension that is to be controlled. There are other arrangements equivalent to this, but this one makes the action the clearest.

This physical arrangement of components is further constrained by the requirement that the system always oppose disturbances tending to create a nonzero error signal; this is tantamount to saying that the system must be organized for negative (not positive) feedback, and that it must be dynamically stable—it must not itself create errors that keep it "hunting" about the final steady-state condition. There is no point in concern with unstable systems, because the (normal) behavior we wish to explain does not show the symptoms of dynamic

Living Control Systems

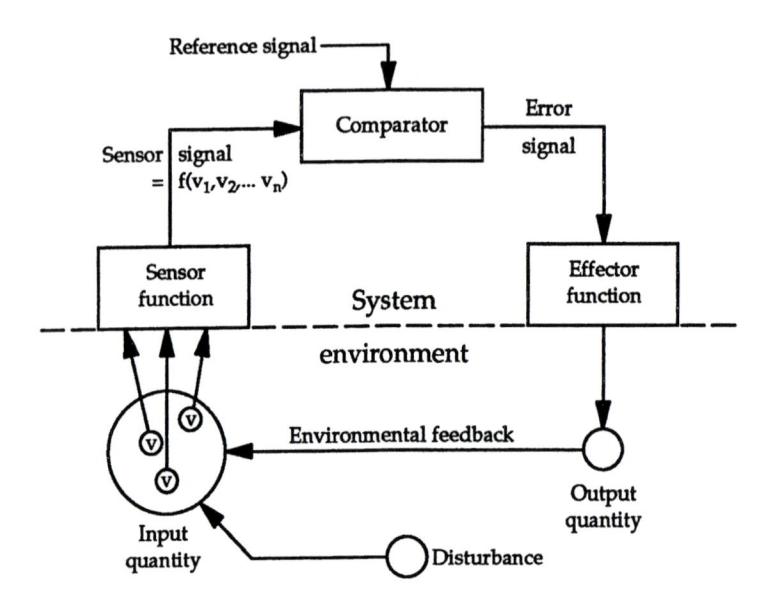

Fig. 1. Basic control-system unit of behavioral organization. The *Sensor function* creates an ongoing relationship between some set of environmental physical variables (v's) and a *Sensor signal* inside the system, an internal analog of some external state of affairs. The sensor signal is compared with (subtracted from, in the simplest case) a *Reference signal* of unspecified origin (see text). The discrepancy in the form of an *Error signal* activates the *Effector function* (for example, a muscle, limb, or subsystem) which in turn produces observable effects in the environment, the *Output quantity*. This quantity is a "response" measure. The environment provides a feedback link from the output quantity to the *Input quantity*, the set of "v's" monitored by the sensor function. The input quantity is also subject, in general, to effects independent of the system's outputs; these are shown as a *Disturbance*, also linked to the input quantity by environmental properties. The disturbance corresponds to "stimulus." The system, above the dashed line, is organized normally so as to maintain the sensor signal at all times nearly equal to the reference signal, even a changing reference signal. In doing so it produces whatever output is required to prevent disturbances from affecting the sensor signal materially. Thus the output quantity becomes primarily a function of the disturbance, while the sensor signal and input quantity become primarily a function of the reference signal originated inside the system. For all systems organized in this way, the "response" to a "stimulus" can be predicted if the stabilized state of the input quantity is known; the stimulus-response law is then a function of environmental properties and scarcely at all of system properties.

instability—and we do not have to design the system.

This system is modeled after Wiener's original concept (6). In the system I describe, however, there are certain changes in geometry, particularly the placement of the system boundary and the identification of the sensor (not reference) signal as the immediate consequence of a stimulus input. This is a continuous-variable (analog) model, without provision for learning.

A system that meets these requirements behaves in a basically simple way, despite the complexities of design that may be required in order to achieve stable operation. It produces whatever output is required in order to cancel the effects of disturbances on the signal generated by the sensor. If the properties of the sensor remain constant, as we may usually assume, the result is to protect the controlled quantity against the effects of unpredictable disturbances of almost any origin.

Goal-Directed Behavior

The reference signal constitutes an explanation of how a goal can be determined by physical means. The reference signal is a model inside the behaving system against which the sensor signal is compared; behavior is always such as to keep the sensor signal close to the setting of this reference signal.

With this model we gain a new insight into so-called "goal-seeking" behavior. The usual concept of a goal [for example, William Ashby's treatment (7)] is something toward which behavior tends over some protracted period of time. We can see that idea now as describing the behavior of a sluggish control system, or a control system immediately after an overwhelming disturbance. Many complex control systems are sluggish, but only because any faster action would lead to dynamic instability. The appearance of "working toward" a goal may result from nothing more than our viewing the system on an inappropriately fast time scale.

It is useful to separate *what* a control system does from *how* it does what it does. Given two control systems controlling the same quantity with respect to the same reference signal, one system might be able to resist disturbances lasting only 0.1

second while the other could not oppose a disturbance lasting less than 1 second. After a disturbance, one system might restore its error signal nearly to zero in one swift move, while the other makes that correction slowly and after several over- and undershoots of the final steady-state condition. These are dynamic differences, and have to do with the details of system design. Both systems, however, do the same thing when viewed on a slow enough time scale: they control a given quantity, opposing disturbances tending to affect that quantity. On a time scale where we can see one system "working toward" the goal state, we might see the other as never allowing significant error to occur—as reacting simultaneously with the disturbance to cancel its effects.

The proper time scale for observing what a control system does is that on which the response to an impulse-disturbance is apparently zero. That automatically restricts our observations of disturbances in the same way: all disturbances appear to be slow. On such a slow time scale, it is apparent that a control system keeps its sensor signal nearly matching its reference signal by producing outputs equal and opposite to disturbances, in terms of effects on the controlled quantity.

The normal behavior of a good control system, viewed on the appropriate time scale, is therefore not goal-seeking behavior but goal-maintaining behavior. The sensor signal is maintained in a particular goal state as long as the system is operating within its normal range, in the environment to which its organization is matched. If the properties of the sensor do not change, this control action results in the external controlled quantity being maintained in a state we may term its reference level.

Much of what we interpret as a long process of goal-seeking (and perhaps all) can be shown to result from higher-order goal maintenance that involves a program of shifting lower-order reference levels, but that anticipates what has yet to be developed here.

Controlled Quantities

The key concept in this model, as far as observable behavior is concerned, is that of the controlled quantity. If it were possible to identify a controlled quantity and its apparent reference level, the model just given would provide an adequate physical explanation for existence of this quantity and its goal state, just as the telephone-switchboard model of the brain has heretofore been taken as an adequate physical explanation for stimulus-response phenomena. To be sure, the source of the reference signal that sets the system's goal remains unspecified, but that is of no consequence in a part-model of a specific behavior pattern. We are concerned here with immediate causation, not ultimate causes.

If a quantity is under feedback control by some control system, that fact can be discovered by a simple (in principle) procedure, based on the fact that the system will oppose disturbances of the controlled quantity.

Suppose we can observe the immediate environment of a control system in terms of detailed physical variables (v_1, v_2,... v_n). We postulate a controlled quantity $q_c = f(v_1, v_2,... v_n)$, where f is a function of the variables. According to the definition and known physical principles, we can then devise a small disturbance d affecting some v's such that (in the absence of behavioral effects) $\Delta q_c = g(d)$, where g is the function describing the environmental connection between the disturbance and the controlled quantity. Applying the disturbance we predict a change in q_c, and compare it with the observed change, $\Delta^* q_c$. If we have hit upon a definition of q_c that is accurate, and if a reasonably good control system is acting, we will find $\Delta^* q_c / \Delta q_c \ll 1$.

By progressively changing the definition of q_c [that is, the form of f in $f(v_1, v_2,... v_n)$], we can find a minimum in the ratio $\Delta^* q_c / \Delta q_c$; that is, we can find a definition of the controlled quantity such that the observed effect of a disturbance is far less than the effect predicted according to physical principles, omitting the behavior of the system.

The reason for the "failure" of the prediction is of course the fact that the control system actively opposes effects of d on q_c. Let h be the function describing the environmental connection between the output o of the system and the controlled quantity. If the output o affects q_c additively according to the relationship $\Delta q_c = h(o)$, then the total effect on q_c is the sum of the effects of the disturbance and the system's active output: $\Delta q_c \cong g(d) + h(o)$. When control is good, this sum will be nearly zero.

Defining the zero points of the controlled quantity and the system's output as their undisturbed values, we can see that the controlled quantity will remain nearly at its zero point ($\Delta q_c \approx 0$), while the disturbance and the system's output will be related by the approximation, $g(d) \approx -h(o)$.

Here is a very simple example. Suppose we observe a soldier at attention, and guess that one controlled quantity involved in his behavior is the vertical orientation of one of his arms, seemingly being held in a straight-down position (the zero point). If this quantity were not under active control, we could predict that a sideways force of 1 kilogram would raise the arm to about a 30-degree angle from the vertical. Applying the force, we observe that in fact the arm moves only 1 degree, or 1/30 of the predicted amount. The effective force-output of the soldier is thus just a trifle under 1 kilogram in a direction opposite to our 1-kilogram disturbance, the trifle being the restoring force due to the slightly deflected mass of the arm, and gravity. This is a reasonable verification of the initial guess, and we may claim to have found a control system in the soldier by identifying its controlled quantity.

The reference level of a controlled quantity can better be defined as its value when the system's output is totally unopposed (even by friction or gravity). Because that state normally implies no error-correcting output, the reference level of the controlled quantity can also be defined as that level (state, for multidimensional quantities) which results in zero error-correcting output.

A controlled quantity need not have a reference level of zero. The soldier, for example, might be persuaded to raise his arm to the horizontal position, so that in the same coordinate system used before, the apparent reference position is now 90 degrees. The weight of the arm now constitutes a natural disturbance, and we would guess that the system's output is now equivalent to an upward force equal to the weight of the arm. If that force were 10 kilograms, we would also predict that an upward force disturbance of 10 kilograms would cause the arm muscles to relax completely, or at least that the net force-output would drop to zero (arm muscles can oppose one another). Our pushing upward with a force of 11 kilograms should result in an output of 1/2 kilogram downward.

Hierarchies of Controlled Quantities

Suppose that the soldier is now ordered to point at a passing helicopter. He will raise his arm and do so. We can verify that arm position is still a controlled quantity by applying force-disturbances, but now the picture is complicated. The test still works for relatively brief (but not too brief) disturbances, but over a period of some seconds we find that arm position does not remain constant. Instead, it moves slowly and uniformly upward and sideways, as the helicopter approaches.

This suggests that a second controlled quantity has entered the picture. If the helicopter stops and hovers, this new controlled quantity is invisible—the force-test cannot distinguish it, for the arm simply remains almost still as before. But if we radio the helicopter pilot to move his craft in various ways, we can test the hypothesis that the soldier is controlling the angular deviation of his pointing direction from his actual line of sight to the helicopter. If that were not a controlled quantity, the pilot's moving the helicopter would create a predictable deviation. In fact, movement of the helicopter results in no observable deviation at all (barring slight tremors). We are reasonably assured that the pointing direction relative to the direction of the helicopter (and nothing else) is a two-dimensional controlled quantity, with a reference level of zero deviation.

Now we have a slight dilemma. We established, and could reestablish at any time, arm position as a controlled quantity. (The position-control system will react to disturbances within the lag time of the pointing-control system.) Yet control of the new controlled quantity requires a change in arm position, which would constitute a disturbance of the first system. Why does the first control system not resist this change?

The answer is obvious. The second control system opposes disturbances not by direct activation of force outputs, but by altering the reference level, by means of changing the reference signal for the arm-position control system.

Now two controlled quantities (and implied control systems) exist in a relationship that is clearly hierarchical. One controlled quantity is controlled by means of changing the reference level with respect to which a second quantity is controlled. This immediately suggests a partial answer to the question raised by Fig. 1: Where does the reference signal come from? It is clearly the output of a higher-order control system, a system that senses a different kind of quantity and controls it with respect to an appropriate reference signal by using the whole lower-order system as its means of error prevention (the appropriate time scale for the higher-order system will be slower than that for the lower).

We now have a plausible physical model for a two-level structure of goals. The goal of pointing is achieved by setting —and altering—a goal for arm position. In fact the higher-order system must adjust reference levels for two lower-order control systems, one governing horizontal arm position and one governing vertical arm position: both can be shown to be under feedback control. Of course we do not know yet the actual nature of the lower-order systems—any two non-collinear directions of control would give the same observed results. But we have achieved a first approximation.

The source of the lower-level reference signals has been identified but the question of the ultimate source of reference signals has simply been pushed up a level. The range of explanation for immediate causes, however, has been considerably extended.

This hierarchical analysis of behavior can now be continued indefinitely, the only restriction on the number of levels being that imposed by experimental findings. The model of the brain's organization (for that is what it is) can be extended accordingly. Each time a new level of control is found, the range of explanations of immediate causes of behavior is extended to cover more kinds of behavior and to span longer periods of time. Each such extension redefines the question of ultimate causes, for each new level of reference signals represents goals of greater generality.

Our going up a level in this analysis is equivalent to our asking what purpose is served by achievement of a given set of lower-order goals: *why* is the man doing that? Why does the soldier raise his arm? In order to point at the helicopter. Why does he point at the helicopter? Perhaps—we would have to verify this guess by test—perhaps to comply with an order. And why comply with an order...?

Going down a level is equivalent of asking *how*. How must the man behave in order to point? He must control his arm position. How must he behave in order to control arm position? He must control net muscle-generated forces. And the chain extends further down, to the control systems in the spine which control the effort in whole muscles, as sensed kinesthetically. Each level must be verified by finding a way of disturbing the controlled quantity without affecting lower-order quantities.

Oddly enough, behaviorists may have already found the answer to the ultimate *why* at the top level of the model. Why are the highest-order behavioral goals set where they are set? In order to control certain biologically important variables, which Ashby called critical variables and which I term intrinsic quantities. These are the quantities affected by deprivation and subsequent reinforcements that erase, or at least diminish, the errors caused by deprivation. This makes the highest order of reference levels into those extremely generalized ones that are inherited as the basic conditions for survival. But that takes us to the verge of learning theory, which is beyond the intent of this article. Briefly, I view the process of reorganization itself as the error-driven "output" of a basic inherited control system which is ultimately responsible for the particular structure of an adult's behavioral control system (8). For a human being, the "intrinsic reference levels" probably specify far more than mere food or water intake. We cannot arbitrarily rule out any goal at this level—not even goals such as "self-actualization."

Implications for Behaviorism

The most important implication of this analysis for the traditional view of cause and effect in behavior lies in the fact that control systems control what they *sense*, not really what they

74   Living Control Systems

do. In the total absence of disturbances, a control system hardly needs to do anything in order to keep a controlled quantity at a reference level, even a changing reference level. By far the largest portion of output effort is reserved for opposing disturbances.

This is expressed in the approximate relationship, $g(d) \cong -h(o)$. Because of the way negative feedback control systems are organized, the system's output is caused to vary in almost exact opposition to the effects of disturbances—the chief determinant of output is thus the disturbance. If we read "stimulus" for disturbance and "response" for some measure of output, stimulus-response phenomena fall into place within the feedback model.

Stimuli do cause responses. If one knew the controlled quantity associated with a given stimulus-response pair, one would see more regularity in the relationship, not less. In fact one would see an exact quantitative relationship, for the effects of the response on the controlled quantity must come close to canceling the effects of the stimulus on that same quantity, and both these effects are mediated through the environment, where the detailed physical relationships can be seen. That implies, of course, that given knowledge of the controlled quantity one can deduce the form of stimulus-response relationships from physical, not behavioral laws (9).

Knowledge of the controlled quantity makes the stimulus-response relationship even clearer by pointing out the right response measure and the right measure of the disturbance, or stimulus. An organism's muscle efforts produce many consequent effects, no one of which can be chosen on the basis of behavioristic principles as being a "better" measure than any other. A stimulus event impinges on an organism and its surroundings in many ways and via many paths, again undistinguishable under the philosophy of behaviorism. Knowledge of the controlled quantity eliminates irrelevant measures of stimulus and response.

Let us consider a rat in a Skinner box. The rat responds to a light by pressing a lever for food. Whatever the immediate controlled quantity may be, it is clearly not affected by the current that flows to the apparatus when the lever is depressed: opening the circuit will not in any way alter the rat's next press of the lever. But holding the rat back with a drag-harness as it moves toward the lever would create immediate forward-pushing efforts, so we would know that the rat's "motion" is close to a controlled quantity. We would of course try to do better than that.

Even though the current to the experimental apparatus does affect the appearance of food, which is quite likely to be a controlled quantity (q_c), the current is still not a controlled quantity, for we could leave the circuit open and actuate the food dispenser in a different way, and the rat would still do nothing in opposition, nothing to restore the current. There is no need to assume what is controlled except as a starting hypothesis, and this method can disprove wrong hypotheses.

The irrelevance of some stimulus measures is common knowledge; rats, for instance, have been found to respond quite well to a burned-out stimulus light, provided that the actuating relay still clicked loudly enough.

Systematic experimental definition of controlled quantities will eliminate irrelevant side effects of stimuli and responses from consideration. But it will also negate the significance of most stimulus-response laws, for once a controlled quantity has been identified reasonably well, a whole family of stimulus-response laws becomes trivially predictable. Once it is known why a given response follows a given stimulus, further examples become redundant. Knowing why means knowing what is being controlled, and knowing the reference level.

When a controlled quantity is found, variability of behavior is drastically lowered, simply because one no longer considers irrelevant details. The remaining variability is lowered even further as one explores the hierarchy of controlled quantities. If all we observed about the soldier in the example were his force outputs, we would have to fall back on statistics to predict them. If we then understood that the soldier was using these outputs to control arm position we could find many cases in which there would be scarcely any variability; applying the correct stimuli (forces) would result in quantitatively predictable force outputs. There would still be many unpredicted changes, but a good fraction of those would become precisely predictable if we understood that the soldier was using arm position in order to point at a specific moving object. Of course as we push toward higher and higher orders of control organization we will find more complex systems employing many lower-order systems at once so that prediction depends on our determining which of several apparently equivalent subsystems will be employed. In principle, however, we can become as thoroughly acquainted with one individual's structure of controlled quantities as we please, if cooperation continues to satisfy his higher-order goals.

Control systems, or organisms, control what they sense. The application of a disturbing stimulus does not affect for long what matters to the organism at the same level as the disturbance, because the organism will alter its lower-order goals in such a way as to cancel the effects of the disturbance. If a position disturbance is applied, the organism will alter its force goals and prevent disturbance of position. If a relative position disturbance (movement of the helicopter) is applied, the organism will alter its absolute position goals and prevent disturbance of relative position.

In this way the system continues to oppose disturbances, making adjustments at every level in the hierarchy of control. The organism will not let *you* (the experimenter) alter what it senses (if it can prevent it), but it will without hesitation alter the very same quantity itself in order to prevent the experimenter's disturbing a higher-order controlled quantity. Hence the well known perversity of experimental subjects!

It is this hierarchical character of control systems that makes it seem that organisms value self-determinism. And that is not only appearance: organisms are self-determined in terms of inner control of what they sense, at every level of organization except the highest level.

Only overwhelming force or insuperable obstacles can cause an organism to give up control of what it senses, and that is true at every level. In order to achieve ultimate control over behavior, one must obtain the power to deprive the organism of something its genes tell it it must have, and make restoration contingent on the organism's setting particular goals in the hierarchy of learned systems, or even on acquiring new control systems. But one attempts that at risk. Human beings are more prone to learn how to circumvent arbitrary deprivation than they are to knuckle under and do what someone else demands in order to correct intrinsic error. In the sequence deprive, reward, deprive, reward..., one person may see the reward as terminating deprivation, but that is only a matter of perceptual grouping. Another person may learn that reward leads to deprivation, and take appropriate action against the cause of deprivation. Pigeons in Skinner boxes, of course, do not have that option.

Summary

Consistent behavior patterns are created by variable acts, and generally repeat only because detailed acts change. The accepted explanation of this paradox, that "cues" cause the changes, is irrelevant; it is unsupported by evidence, and incapable of dealing with novel situations.

The apparent purposefulness of variations of behavioral acts can be accepted as fact in the framework of a control-system model of behavior. A control system, properly organized for its environment, will produce whatever output is required in order to achieve a constant sensed result, even in the presence of unpredictable disturbances. A control-system model of the brain provides a physical explanation for the existence of goals or purposes, and shows that behavior is the control of input, not output.

A systematic investigation of controlled quantities can reveal an organism's structure of control systems. The structure is hierarchical, in that some quantities are controlled as the means for controlling higher-order quantities. The output of a higher-order system is not a muscle force, but a reference level (variable) for a lower-order controlled quantity. The highest-order reference levels are inherited and are associated with the meta-behavior termed reorganization.

When controlled quantities are discovered, the related stimulus-response laws become trivially predictable. Variability of behavior all but disappears once controlled quantities are known. Behavior itself is seen in terms of this model to be self-determined in a specific and highly significant sense that calls into serious doubt the ultimate feasibility of operant conditioning of human beings by other human beings.

References and Notes

1. L. von Bertalanffy, in *Toward Unification in Psychology*, J.R. Royce, Ed. (Univ. of Toronto Press, Toronto, 1970), p. 40.
2. W.T. Powers, *Behavior: The Control of Perception* (Aldine, Chicago, in press).
3. This distinction is akin to the older distinction between movement and action, the more recent distinctions between molecular and molar, or proximal and distal aspects of behavior. What I term an *act* is a behavior that is arbitrarily left unanalyzed, while a *result* is defined as an understandable physical consequence of an act. Act and result are relative terms, whereas those they replace are absolute. In some circumstances it may be appropriate to consider a movement as a result, in which case the *acts* would be the tensing of muscles. What is proximal or molecular at one level of analysis may be distal or molar at another level. "Distal achievement," in this feedback theory, becomes *perceptual* achievement, and is multiordinate.
4. W.W. Rozeboom, in *Toward Unification in Psychology*, J.R. Royce, Ed. (Univ. of Toronto Press, Toronto, 1970), p. 141.
5. E. Brunswik, *The Conceptual Framework of Psychology* (Univ. of Chicago Press, Chicago, 1952).
6. N. Wiener, *Cybernetics: Control and Communication in the Animal and the Machine* (Wiley, New York, 1948).
7. W.R. Ashby, *Design for a Brain* (Wiley, New York, 1952).
8. W.T. Powers, R.K. Clark, R.L. McFarland, "A General Feedback Theory of Human Behavior," in *General Systems—Yearbook of the Society for General Systems Research 1960*, L. von Bertalanffy, Ed. (Society for General Systems Research, Ann Arbor, Mich., 1960), pp. 63-83.
9. W.T. Powers, *Behav. Sci.* 16, 588 (1971).
10. This article is adapted from a series of lectures given at a faculty seminar on "Foundations of Science," held at Northwestern University, 1971.

[1973, with William M. Baum and Hayne W. Reese]

Behaviorism and Feedback Control

Although there is much of value in the article "Feedback: Beyond behaviorism" by W.T. Powers (26 Jan., p. 351), it is based on an outdated and misconceived idea of behaviorism.

Behaviorism consists in the view that a scientific psychology must deal with the observable. From this proposition, it follows that psychology should be a science of behavior, and that explanations of observed phenomena should be couched in the same terms as the observations themselves, rather than invoking imagined autonomous entities ("explanatory fictions") as causes. Many, perhaps most, psychologists today are behaviorists.

Since its points are mainly methodological, behaviorism never has been wedded to any particular conception of behavior. Early behaviorists perhaps held views similar to the one Powers criticizes, but the inadequacy of describing behavior in terms of responses to stimuli was recognized over 30 years ago. With the recognition that behavior is affected by its consequences (the Law of Effect), open-loop descriptions began to pass away. Few behaviorists today would disagree with Powers's statement, "there can be no nontrivial description of responses to stimuli that leaves out purposes." Emphasis on purpose, in fact, has been the hallmark of modern behaviorists' thinking (1). The behaviorists' solution to the problem of purpose has been exactly the one suggested by Powers—selection by consequences. That behavior and consequences constitute a feedback system is taken as a basic premise (2). It is presented

Copyright 1973 by the AAAS. Reprinted with permission of the publisher from *Science* 181(4105), September 21, 1973, 1114, 1116, 1118-1120.

this way in at least one elementary text (3).

Powers covers familiar ground in two other points. In his discussion of acts and results, he actually reinvents Skinner's concept of the operant (4). One of Skinner's most important innovations was this conception of a unit of behavior consisting of the class of responses (Powers's "acts") defined by its environmental effect (Powers's "result"). As Herrnstein has pointed out (1), Skinner's approach to the problem of purpose was to define behavior in terms of its consequences. Also familiar is the notion of the hierarchical organization of behavior. Lashley (5) made the earliest clear statement of this view. He argued, as does Powers, for a hierarchy of goals and subgoals in behavior. It seemed the only way to account for organized sequences.

Although Powers's attack on behaviorism is misguided, and many of his ideas have been set down before, nevertheless the constructive aspects of the article deserve praise. The very lack of novelty itself shows that Powers, albeit unwittingly, is square in the mainstream of modern behaviorists' thinking about instrumental behavior. His discussion of feedback, therefore, is most welcome, because it helps define the direction in which we are moving.

WILLIAM M. BAUM

Department of Psychology,
Harvard University,
Cambridge, Massachusetts 02138

References and Notes

1. R.J. Herrnstein, introduction to J.B. Watson, *Behavior* (Holt, Rinehart, and Winston, New York, 1967).
2. The opening sentence of *Schedules of Reinforcement* by C.B. Ferster and B.F. Skinner (Appleton-Century-Crofts, New York, 1957) reads, "When an organism acts upon the environment in which it lives, it changes that environment in ways which often affect the organism itself."
3. H. Rachlin, *Introduction to Modern Behaviorism* (Freeman, San Francisco, 1970). See also D.J. McFarland, *Feedback Mechanisms in Animal Behaviour* (Academic Press, London, 1971) and P. van Sommers, *The Biology of Behaviour* (Wiley, Sydney, 1972).
4. B.F. Skinner, *The Behavior of Organisms* (Appleton-Century-Crofts, New York, 1938). See J.R. Millenson, *Principles of Behavioral Analysis* (Macmillan, New York, 1967) for a treatment in terms of set theory.

5. K.S. Lashley, in *Cerebral Mechanisms in Behavior*, L.A. Jeffress, Ed. (Wiley, New York, 1951), p. 112.

Powers briefly describes a closed-loop feedback model of behavior, with special reference to purposive behavior. The model is of interest and deserves serious consideration as an alternative to other behavioral models, but there are some points about the presentation that warrant critical comment.

First, as a model, the system can do no more than *represent* the phenomena in the domain encompassed. A model (of the type under consideration here) provides no explanations, except in the sense of intuition or analogy. Powers does not describe the theory to be associated with the model, and therefore no real explanations are provided.

Second, Powers asserts that no behavioristic model has been able to account for purpose; but in fact purpose has been adequately derived from such behavioristic constructs as the conditioned goal response (the fractional anticipatory goal response, r_g) and other mediational response. In Powers's system, "purpose" is like a template; its effect is not goal-seeking behavior but goal-maintaining behavior, and it is concurrently represented in the system. Powers does not provide adequate, empirically based definitions of the key concepts, such as "reference signal," and in this sense his model is nonbehavioristic. Nevertheless, as far as one can determine, the model is mechanistic, in that the components of the feedback loop are analyzed as a unidirectional, linear causal chain. The very fact that the components can be analyzed in this way indicates that there is no dialectic interpenetration, or reciprocal interaction, because in such interactions the components are inseparable from the whole or structure that comprises them (1).

Powers concludes that "Behavior itself is seen in terms of this model to be self-determined in a specific and highly significant sense that calls into serious doubt the ultimate feasibility of operant conditioning of human beings." Were it not for the ambiguity of the meaning of "ultimate feasibility," one could reject the statement on empirical grounds. The research literature is replete with studies demonstrating operant conditioning in human subjects by human experimenters, in some cases without the subjects' being aware even that they were subjects (2). One can also, in any case, question the statement on theoretical grounds, because "self-determined" has, as Powers says, a specific meaning in the model, and this meaning has no implication of free will. In the model, "organisms are self-determined in terms of inner control of what they sense, at every level of organization except the highest level." Inner control refers to feedback ("error signal") regarding the discrepancy between the "reference signal," or goal, and the "sensor signal," or input. As the model is mechanistic, the error signal will inexorably produce specific "effector functions," or responses. That is, the responses are completely determined by the error signal (and, perhaps, by the state of the system), and the error signal is completely determined by the reference signal and sensor signal. As the sensor signal is determined by the environment, any variability in self-determination must come from variability in reference signals. Their source is not specified in the model (except at the highest level, at which they are assumed to be biogenetically determined). The model does not demand a reference signal that prohibits operant conditioning; this reference signal was introduced ex hypothesi and is not entailed by the model.

In summary, the model (i) is not explanatory, (ii) is not the only mechanistic model that provides a derivation of purpose, and (iii) does not intrinsically preclude human operant conditioning.

HAYNE W. REESE

Department of Psychology,
West Virginia University,
Morgantown 26506

References

1. H.W. Reese and W.F. Overton, in *Life-Span Developmental Psychology: Research and Theory*, L.R. Goulet and P.B. Baltes, Eds. (Academic Press, New York, 1970), pp. 115-145; W.F. Overton and H.W. Reese, in *Life-Span Developmental Psychology: Methodological Issues*, J.R. Nesselroade and H.W. Reese, Eds. (Academic Press, New York, 1973), pp. 65-86; W.F. Overton, *Hum. Develop.*, in press; L. von Bertalanffy, *General System Theory* (Braziller, New York, 1968).

2. H.M Rosenfeld and D.M. Baer, *Psychol. Rev.* 76, 425 (1969).

The comments by Baum and Reese on my control-system approach to understanding behavior are the most balanced I have received from behaviorists. I thank them for trying to find a place for my work within behaviorism, an attempt that reflects generosity, but not understanding, of what I said (or tried to say). The conceptual basis of control-system theory is so alien to behavioristic thought that there can be no such easy reconciliation. The best we can hope for is a constructive confrontation.

Baum says that a scientific psychology must deal with the observable, which to him means behavior. Behavior, however, is not something self-evident that anyone can see just by looking. What is the behavior of a man walking? Is he really tensing his leg muscles, moving his legs, walking, going to buy a paper, on his way to work, making a living for his family, or maintaining his self-respect? The point of view of the observer defines the behavior he sees. The *actual* behavior of the nervous system consists only of sending neural signals to muscles and glands; that is the last event that truly reflects the system's output. From that point outward, the results of that output become more and more mixed with properties of any events in the external physical environment, so that even such elementary behavior as a "movement" no longer is a unique indicator of a particular activity in the nervous system. Thus, while Baum's pronouncement seems reasonable on the surface, it ignores one of the deepest conceptual dilemmas of behaviorism.

The control-system model shows that behavior at any level, as well as its relationship to "stimulus events," makes sense as soon as one recognizes the concept of the controlled quantity. To find the proper definition of the controlled quantity, the observer must recognize that his own point of view determines the behavior he will observe, and he must find an objective way to discover the *right* point of view—namely, that of the behaving system. The observer must try to find out which of the infinity of potential controlled quantities is the one that the behaving system is actually sensing and controlling. Only when the controlled quantity has been correctly identified can the observer see that the system's outputs are always such as to counter the effects which environmental disturbances would otherwise have on the controlled quantity. In my article I presented an experimental paradigm, new to psychology, for testing hypotheses concerning the controlled quantity and its reference level.

In the section on controlled quantities in my article, there appears an approximation, $g(d) \approx -h(o)$, which says that the cause-effect relationships that can be observed between stimulus events and consequences of nervous system outputs—responses—are expressible wholly in terms of the physics of the local environment, containing almost no information about the behaving system at all. I see no way in which behaviorism can survive a full understanding of the derivation and significance of this harmless expression. If control-system theory does indeed describe correctly the relationship between organisms and their environments, behaviorism has been in the grip of a powerful illusion since its conceptual bases were laid.

It is therefore not possible that behaviorism already contains an adequate treatment of feedback phenomena; if it did, a behaviorist would have discovered this illusion already. Many behaviorists have observed feedback phenomena, but they have tried to deal with them by translating the terminology of control-system theory in such a way that well-accepted behavioristic principles would remain undisturbed. That is why "purpose" has lost its original meaning of inner purpose or intentionality, and has been redefined as *consequences*. That redefinition was necessitated by the fact that early behaviorists knew of no physical system that could contain inner purposes —their telephone-switchboard model had no place for them, and control-system theory then lay far in the unforeseeable future.

In control-system terms, a purpose is not a consequence of behavior, but a model inside the organism for what it wants the perceptual consequences of its outputs (modified by environmental disturbances or not) to be. When I bowl, my inner purpose is to perceive all the pins falling on the first ball. What I perceive is generally something different. I am still doing my best to alter my outputs in such a way as to reduce the error between what I generally perceive and what I intend to perceive. Another observer can discover that intended perception by manipulating my environment until he finds the state where I cease to alter my outputs in opposition to the changes he causes. There is nothing metaphysical or conjectural about this process. But it does not make any sense in behavioristic terms, because it is designed around rigorous laws of feedback, not around the imprecise usages of the term feedback that are found in behaviorism.

There seems to be a general impression that feedback is analyzable (in Reese's terms) by following a "unidirectional, linear causal chain" around and around a closed circle (I trust that Reese noticed that the circle *is* closed). That approach to feedback, often expressed as taking into account the effects of a response on subsequent stimuli, is the natural one, but, as every beginning control-system engineer soon discovers, it leads to totally incorrect predictions of the behavior of the system being modeled. The qualitative chain-of-events approach leaves out the crucial factor of system dynamics; when that is properly taken into account, through use of a physical analysis of the system and its environment and application of differential equations or transform methods, a very different and surprising picture emerges. If the control system one wants to model is free of spontaneous, self-sustained oscillations (as normal behavioral systems are), time lags in the system can safely be ignored, and the behavior of the whole system can be seen quite correctly as occurring *simultaneously* with disturbances. The output changes *along with* the disturbance (a normal, slowly varying disturbance), and the input variable being monitored continually tracks the inner reference signal, if a variable inner reference signal exists. There are no loopholes in this analysis; if organisms are in the negative feedback relationship with their environments, this is how they behave. To arrive at a different conclusion, one would have to show that the bases of control-system theory are wrong, and he would have a lot of engineers who use it every day to convince.

Thus, the attempts by behaviorists to bring feedback phenomena into the scope of their conceptual scheme have involved only a superficial adoption of certain terms and loose qualitative observations, the true beauty and power of control-system concepts having been left behind. The distortions of feedback theory that occurred in the adoption of the terminology were precisely those which would prevent change in the basic conceptual scheme of behaviorism (this should not surprise control-theory fans, since all organisms manipulate their own perceptions to keep them in the desired state).

There is no "reference signal that prohibits operant conditioning," as Reese puts it while guessing wrong about what I meant. Operant conditioning is a fact; in my model, it is a portion of a control process whereby organisms modify their own inner structure of control systems as a means of keeping certain critical variables (W.R. Ashby's term, as I noted), at genetically established reference levels. I was talking about the *feasibility* of people deliberately trying to control the behavior of other people through deliberate application of operant conditioning.

In order to control another person, one must establish contingencies or schedules of reinforcement. Whatever one chooses to use as a reward, he must make sure (i) that the subject needs or wants the reward and (ii) that the *only* way the subject can obtain the reward is by doing what the experimenter wants to perceive him doing. The experimenter, of course, is trying to control his own perceptions relative to his own inner purposes, using the subject as his means.

The establishment of contingencies, therefore, requires that the experimenter already be the sole source of something the subject wants, and establishing that situation is where operant conditioning will fail as a way of controlling behavior—as it has failed throughout recorded history. An experimenter trying to control people rather than laboratory animals cannot conceal the fact that he has what the subject wants, and is withholding it until the subject does what the experimenter demands. If one person can establish a contingency, another person can see that he has done so, and can decide to "unestablish" it. If the act that the experimenter wants to see performed in any way inconveniences the subject, the subject will be forced by his own nature to find a way to circumvent the contingency. He can operate properly only on the basis of his own inner purposes, not on the basis of the experimenter's.

Only a god capable of seeing a person's entire structure of inner goals could establish contingencies for that person without creating conflicts that would lead to a direct and violent confrontation. Even then, the god would be constrained to controlling the person in ways that created no uncorrectable errors in that person's control hierarchy.

Operant conditioning is only a modern term for what people have been trying to do to each other since civilization started. Everyone knows that people seek rewards and will change their behavior, within limits and as necessary, to get those rewards. But rewarding always implies withholding, and withholding what people need is a sure way to create violent and bloody conflict. An excellent case can be made for the statement that the present state of the world is the direct result of people trying to set up contingencies of reward for each other. It is time we realized that this principle of social interaction is the cause of, not the solution to, our most serious human problems.

Finally, I want to acknowledge the justice of some of the criticisms of my work. I know that I have overgeneralized in speaking of "behaviorists" when I really should have said, "some behaviorists." My aim is to find ways to effect a transition from what I believe is an outmoded view of the nature of human nature—and animal nature—to what seems a vastly more productive and humane point of view. My attitude toward what I see as the basic errors of behaviorism is not one of irritation or superiority. My model is only a feeble step in the right general direction. It is instead that there is an enormously difficult task ahead—but, considering what I see as the possible results of success, worth all the effort. I hope that Baum and Reese and other behaviorists will come to see it this way after careful consideration. I know their task is harder than mine, and it would be even harder if this clash of ideas were set up so that someone had to win, and someone had to lose.

WILLIAM T. POWERS

1138 Whitfield Road,
Northbrook, Illinois 60062

[1974]
Applied Epistemology

Philosophical investigations sometimes seem divorced from the hard realities of science, which force one to take a position in order to get on with the work at hand. The engineer or physical scientist not engaged in basic research, for example, is almost forced into being a naive realist; one cannot build, design, or analyze an electronic or mechanical device while wondering if the soldering iron, meter, or slide rule is really there.

Once in a while, however, a study that began as a comfortable application of the known to the less well known leads to a philosophical impasse, and one finds himself forced to shift philosophical gears. Perhaps it is a sign of some general paradigm shift that many investigators in many disciplines, even in nuclear physics, are experiencing uncertainty about their working premises, and are beginning to realize that naive realism is not the hard-headed practical man's philosophy it once appeared to be. It may, in fact, create an invitation to illusion.

Other segments in this volume deal in detail with the epistemological position of Piaget, showing clearly that this pioneer has been exploring for years a new concept of knowledge. It may be that Piaget has for several decades suffered an extreme of misunderstanding of his position—or it may be that in his direct approach to the growth of perceptual organization, he has been applying new principles without having organized them into an "official" statement. The subject of this chapter is another approach that has converged to the same

Copyright 1974 by William T. Powers. Reprinted from Charles D. Smock and Ernst von Glasersfeld, eds., *Epistemology and Education*, 1974, 84-98.

general conclusions from a totally different starting point: cybernetics. In this case the epistemological principles have emerged in an explicit form simply as a consequence of following out the logic of a behavioral model.

The model on which I have been working is an offshoot of cybernetics using almost the oldest and least sophisticated of cybernetic concepts: feedback of behavioral outputs to sensory inputs, through the environment. Through a rigorous, and some might say obsessive, application of a simple control-system unit of behavioral organization, I have constructed what seems on first inspection to be a purely hardware model of how behavior works, the kind of model that would make any engineer feel secure.

The C-S Unit of Behavior

The basic unit works like this. A given control system (cs) senses some variable outside itself, the state of that variable being represented inside the system as a perceptual signal. The perceptual computer (the "input function") involved may be very simple, as in a spinal reflex, or it may be complex beyond present understanding, as in a system that can sense abstract variables such as the status of a strategy. In any case, the magnitude of the perceptual signal represents the present-time state of some external variable, quantity, condition, situation, and so on: anything perceivable by a human being. One perceptual signal always represents only one external variable, and is one-dimensional.

Also entering this basic organizational unit is a reference signal (I will say in a moment where it comes from). The reference signal has the same physical form as the perceptual signal, probably a train of neural impulses, and, when fixed at a particular magnitude, represents a particular state of the perceptual signal. Comparison of the perceptual and reference signals (by a "comparator," of course), results in creation of an error signal, a neural signal that continually represents the amount and direction of mismatch between the reference and the perceptual signal.

The error signal—amplified, filtered, and copied into many

outgoing channels by the system's "output function"—drives motor behavior.

The most important fact about this model emerges when the feedback path through the external world is introduced. The motor behavior driven by the error signal affects the state of the very external variable that is one of the determinants of the sign[1] and magnitude of the error signal. There is a closed circle of cause and effect.

The net result of putting together these abstract components is a system that behaves in a familiar way: it seeks a goal, the goal being represented by the reference signal. In this model, goals are always goals for perception. We act, if we can, to end up perceiving what we intend beforehand to perceive.

The complete model is made of a whole hierarchy of units like this, a many-levelled organization having many control systems at each level. In general a system at one level makes its perception match the reference signal (or goal perception) it receives from higher systems; its means of doing this is to vary the reference signals of lower-level systems. Thus behavior is organized into a structure of goals and subgoals pertaining to perceptions of different types. Each class of perception represents a class of external variables (such as color-sensation, configuration, motion, relationship), and the idea that perceptions somehow *correspond* to external variables is the source of the epistemological problem we will explore. Before taking up that problem, however, we have to establish the general picture of this closed loop of cause and effect, which we call a control system.

As servomechanism engineers discovered in the early 1940s (and as most psychologists have yet to realize), the consequences of closing this cause-effect circle cannot be predicted correctly by tracing a series of events around (and around and around) the loop. That sort of analysis, when done quantitatively, leads to a model that drives itself, by its own logic, into ever more violent oscillation.

The correct analysis for most cases depends on treating all variables as continuous quantities. It is, after all, only a mat-

---

[1] "Sign" refers to the *direction* of the error: the perceptual signal can indicate an excess over the reference signal, or it can indicate an insufficiency. (Editor's note)

ter of personal choice whether one calls a variable a discrete fact or a point on a continuum of change, and the control-system model seems to show that the latter is always the more productive—and predictive—choice.

When a continuous-variable analysis is employed, some very peculiar—peculiar yet familiar—properties of this control organization emerge. First, for the behavior of the model to match real behavior, we find that the sign of the feedback must always be negative.[2] A disturbance applied anywhere in the closed loop must result in a reaction by the system that opposes the disturbance.

An unfortunate mistake was made by psychologists who were trying to translate the terminology of cybernetics into something more familiar; in effect, they made a pun on the word *positive*. Positive feedback is not "good," "supportive," or "firm" feedback: it is error-enhancing feedback rather than error-opposing feedback. Organisms do not react to disturbances with efforts that make the disturbance even more effective; they oppose disturbances. How else could they maintain any recognizable patterns of behavior in a changing environment? Positive-feedback models, on the other hand, do not behave like normal organisms under normal conditions.

Because of the negative feedback, behavior tends to vary in a peculiar way when disturbances occur. The variation in motor activity, however, is not a direct consequence of the disturbance or "stimulus." The physical causal arrow does not run from disturbance to sensory input, from sensory input to motor output, and from motor output to organized behavior. That is the appearance, of course, and it fits the only concept of mechanism known to man before about 1940, but it is only an illusion. The presence of feedback means that the sensory input is the focus of two causal arrows, one from the disturbance and one from the behavioral output itself. The status of the perception involved, inside the system, can be altered either by independent disturbances or by behavior, equally and simultaneously.

The "peculiar way" in which the behavior of a negative-

[2] I.e., the value fed back must be *different* from the value of the reference signal, thus creating a "disturbance." (Editor's note)

feedback control system varies when a disturbance occurs is easy to characterize. No matter how the disturbance (acting alone) might change the perceptual signal, the behavior that accompanies it produces an equal and opposite effect, thus leaving the perceptual signal nearly unchanged. Since all perceptual signals in this model are one-dimensional, all sources of disturbance of a given perception are equivalent, and their effects on perception can be countered by the same error-based output.

An ideal control system will exactly cancel the effects of any disturbance on its perceptual signal, within its capacity to produce opposing output effects. A control system (that is known to be dynamically stable) approaches the ideal as its error sensitivity increases. Error sensitivity is the amount of error needed to cause maximum output; the greater this sensitivity, the less error is required to produce output equal to the magnitude of disturbance, which is the limiting steady-state condition of an ideal system. Thus higher error sensitivity does not imply more output, but smaller error. That is just one of the little surprises in control-system theory.

The biggest surprise, of course, is that control systems control their own perceptions, not their output activities. When an organism reacts to some change in external circumstances, the appearance is that of a direct cause-effect relationship. If organisms are control systems, the reality is that between the external phenomenon and the alteration of behavior is a variable affected by both, a controlled quantity or controlled variable being sensed by the organic control system and being held at some reference level by behavior. Since that quantity does not change (much), it is not very noticeable, but once it is noticed it completely explains why the observed behavior occurred when the external circumstances changed.

As an example, consider this observation. I open a window, and a person sitting by the window gets up and puts on a sweater. The appearance is that the sight and sound of the window being opened triggered off a complex series of responses, in a direct cause-effect way. In control-system terms the relationship is not direct. Opening the window disturbed a controlled quantity, the temperature of the person's skin, and would have disturbed it a lot more if the person had not done something having the opposite effect—put on a sweater. The behavior was protecting a controlled perception against significant disturbance.

Once a controlled perception (or rather, an observable variable postulated to be equivalent to a controlled perception) has been identified, a whole family of cause-effect relationships can be predicted, and explained. Turning up the thermostat may "make" the person take off the sweater; another person might "respond" to the furnace running out of fuel by going to bed. It is practically always necessary to understand the nature of the controlled quantity to make sense of behavior, and people do this even when they don't realize this is what they are doing. Even behaviorists.

Consider walking a dog. Sometimes pulling on the leash seems to stimulate the dog to pull back in a perverse cause-effect sort of way. But the understanding human will realize that the dog's nose is buried in a delicious smell, and the dog is simply keeping that smell nice and strong. Knowing what the dog is really doing, i.e., what he is controlling, one can then predict quantitatively how much the dog will pull against the leash and in what direction. In fact using such carefully applied disturbances is a formal method for determining what a control system is controlling, even a control system that cannot be questioned.

There are many facets to this kind of model of behavior, but we have now established the concepts of interest here. If organisms are indeed organized as control systems, all their behavior is centered around control of their own perceptions, their own inner representations of the external state of affairs. What we observe as overt behavior is merely the means by which organisms maintain this control. Behavioral activities and their effects on the outside world are not the end-product of the nervous system's actions. One cannot really make sense of any behavior until he knows what perception is being controlled by it, and with respect to what reference level. Only then can one see why a given change in behavior results from applying a given disturbance to the local environment or directly to the organism.

Epistemology and Control

An epistemological problem arose when I tried to do something with this model that is not normally done with behavioral models; i.e., when I tried to apply it to all organisms, including the experimenter as well as his subject. If it is a valid model of organisms, one can hardly claim exemption from it and still defend its generality.

The problem arises from asking how an experimenter could establish the fact that his subject is behaving as a control system with respect to some observable variable. The obvious procedure is direct, in that its object is to see whether behavior satisfies the definition of control behavior, control of input through opposition of disturbances of that input.

Unfortunately, one cannot tell by inspection which aspect of the environment or organism-environment relationship is in fact being perceived at the moment by the organism, much less whether it is being controlled. The mere fact that a variable remains constant proves nothing; it might do so with no organism present, or its constancy might be coincidental to the control of some other variable. When starting from scratch, the experimenter has no choice but to hypothesize a controlled variable, and then try to test the hypothesis.

The test, fortunately, is fairly simple in principle, although applying it requires some inventiveness. Once a potential controlled quantity has been selected, one simply applies disturbances directly to it (for instance by varying other quantities contributing to its state) and sees whether or not the organism's behavior opposes or cancels the effects of the disturbance. If the definition is not quite right, there will be ways of disturbing the proposed controlled quantity that succeed just as if the organism were not there. In that case, one alters the definition to exclude dependence on that type of disturbance, and tries again. The procedure is finished when every possible way of disturbing the variable is met by equal and opposite effects solely due to the organism's actions.

Of course one's final definition may include more than one variable under control by the organism; one may be mistakenly unifying variables that are separately controlled by the organism. The final definition might be a transformation of the actual controlled variable, or it might leave out perceptual functions common to observer and subject. Presumably, experience will show up some of these errors sooner or later. One wonders, though: can this test ever reveal them all?

But that is not how epistemology comes in—that would be too obvious and anticlimactic. The subject of epistemology comes in when we recognize the experimenter's actions as another example of control behavior.

In that segment of experience which the experimenter calls his "mental" world, he is aware of a goal-principle, principles being one of the higher-level classes of perception in the nine-level model (Powers, 1973). The principle or generalization he wants to perceive could be stated, "Disturbances of controlled quantities create equal and opposite effects from behavior." Once he had grasped this principle, he would not describe it; it would become a unitary perception, non-verbal.

What he does perceive at this level would at first have to be stated as a different principle: the disturbance he is trying out is only sometimes opposed, or never. By altering his choices of goals for lower-level perceptions—strategies, relationships, events—he finally makes the perceived principle identical to the goal-principle, eliminating the error.

What we mean by mastering an experimental technique is learning *how* to perceive a principle, and how to select lower-order goals in such a way as to diminish the error. Learning is involved in establishing this control organization in working order, but once the experimenter "gets the hang of it," he is not learning any more but simply behaving as a complex control system, controlling the multilevelled perception called, in this case, "the test for the controlled quantity."

When the test procedure has been carried to a successful end, the experimenter will find that his own perception of the controlled quantity is stabilized against disturbances by the subject's behavior. That is the point where he feels that he now "knows what the subject is doing." He has found an effect of the subject's motor activities that remains the same despite drastic and independent changes in the environment situation.

This, of course, is what any observer is doing when he recognizes a "behavior pattern." A study of behavior has never been a study of outputs. Outputs do not repeat; only their consequences do, and those consequences are defined by what the behaving system is perceiving and controlling. The observer's perceptions do not define the behavior of the subject; the subject's do.

We now have a situation in which at least one variable in the environment, the controlled variable, is represented as a perceptual signal in each of two brains, the experimenter's and the subject's. The processes by which the perceptual signals are derived from sensory contact with the environment are at least similar, the more so if the perceptual signals are invariant with respect to the spatial orientation of the observer. Under a large variety of circumstances, these two perceptual signals will covary, or conversely, be stabilized.

Now the question we have been approaching. Of what are these two perceptual signals a function? The naive realist answers, "Why, the controlled variable, of course." But this test does not reveal any controlled variable; it reveals two covariant perceptual signals, but not the thing of which they are both presumably—a big word—a function. We can call in any number of colleagues, teach them the test, and get them to establish their own perceptions of the controlled variable, but numbers make no difference, consensus does not help. We still have no way of knowing what is actually being controlled in that hypothetical reality we think of as existing independently of our experience of it, the reality *between* your experience and mine. We are all experiencing an analogue of that world, in the form of signals in our brains: the atoms of the world of experience are neural signals. The model tells us that both the model and the experiences we are modelling exist inside our brains (it does not tell us if we should believe this).

Even cross-correlations among senses, or the interposition of artificial sensors, does not enable us to cross the barrier. True, we can cause our visual perceptions to change (by exerting muscular efforts) in ways that lead to changes in tactile perceptions: that may reassure us that what we are seeing is "tangible" (which means, perceivable through the tactile sense), but only if we are desperate for reassurance that we experience Reality Itself.

All right, one may say, it may be true that we cannot see through our perceptual transformations to the cause of perceptions. But does not the fact that human beings share such similar classes of perception show that our perceptual analogues are very much like the real world? I thought so, once. But I ceased to think so when I realized that human beings can control perceptions, through very concrete behavior, that cannot possibly have external counterparts. Find one example and you have a thousand.

As everyone knows (all accusations of prejudice are rejected in advance), home-made soup is always too salty. Different soups need careful adjustment of salt-content, so as to create just the right overall taste. This taste is easy to recognize and, by means of a salt-shaker, to control, although unaccountably hard to communicate to the cook. What is being controlled apparently has no objective significance except to another properly organized nervous system. I presume that example was on familiar ground; but one can go further. With practice, one could learn quite skillful control of the flenishness of his environment. Flenishness is defined as the sum of a Fahrenheit thermometer reading and the time indicating on a clock in hours and fractions of hours. Since these are both meter readings, they must be objective measures of something.

Suppose one determines to keep his environment's flenishness at a reference level of 72. This can be accomplished in a number of alternative ways (a fact that helps assure us of the reality of flenishness). One could, for example, manually adjust the clock reading, while watching the thermometer as well, keeping the sum at 72 as the thermometer reading changes. One could, if this seems like fudging, carry the thermometer about while watching the clock. Toward noon it might be necessary to put the thermometer in the refrigerator to keep a flenishness of 72; the distribution of flenishness is not, apparently, uniform in space or time, but then what is? As a third alternative, one could leave the thermometer and clock alone, and control flenishness by manipulating the air-conditioner and furnace with dedicated disregard of comfort.

Exploring the absurd can sharpen one's eye for it. One does not need to invent outlandish controlled quantities in order to find enough examples of perceptions that can be controlled but have no conceivable reality outside the human brain. Consider an architect designing a new house. He wishes to make the drawing attractive: how shall he adjust the attractiveness of the drawing? By changing proportions, by adding shading, by drawing in a few trees, until the attractiveness is at the architect's reference level for it. The client, of course, will prove boorishly insensitive to the obvious attractiveness of the drawing. Is the problem a lack of sufficient attractiveness? Not at all. It is the absence of an external referent for the term.

Our language offers a multitude of easy ways to speak of our perceptions as if they were really where they appear to be —Outside. We can speak of provocative behavior, pleasant surroundings, understandable explanations, valuable books, gigantic people, strong tastes, scintillating conversation, irritating habits, and even "stimulus value." The list is endless, for it reflects the naive realism which was all our ancestors were capable of. The list "reflects" the habits into which even the most dedicated study of epistemology has fallen, habits that would be exceedingly hard to change. It is frustrating to realize that one's very thoughts flow onto paper in a language that asserts what one knows is not the case.

We 20th century brain-modelers-cum-psychologists must, it seems, resign ourselves to a split-brain sort of existence, hoping that language will evolve enough, in time, to save our descendants this incessant feeling of hypocrisy. In the meantime, we can always talk in algebra.

If there is any validity in this control-system approach, we must face up to a very serious charge concerning human experience. While the model does assume that perceptions are functions of something Outside, it offers no assurance that there is a specific external counterpart of *any* human perception, from elementary sensations on up. It is quite consistent with the model to suppose that human perceptions are arbitrary functions of external variables. These functions become organized because of their survival value (here we are not in disagreement with B.F. Skinner), but the criteria defining "survival value" are within the organism, themselves the results of a control activity in which an entire species participates.

One central concern of this volume, learning, has only been mentioned in passing. This primarily is because the epistemological problems we have been looking at are problems that confront us in our roles as examples of a hierarchical system organized about perception. The control hierarchy we have been looking at has fixed properties; I deliberately left out of this performance model any capacity for self-reorganization, partly just to see how far one could go in accounting for behavior without introducing the capacity for change.

After postponing the question of learning until the performance model has been built, one finds that the question has changed. As pointed out in an earlier example, "learning" what another organism is controlling can be reduced to an algorithm, a procedure that can be learned. Surely learning the algorithm and applying it are two different processes! Once we understand just how much of behavior can be brought under the control-hierarchy paradigm, we begin to see that a very large part of what is called learning has two aspects to it: that which is learned, and the process by which it is learned.

What can be learned? Practically everything. The products of the learning process include rationality, strategies, data about the outside world, modes of interpretation, principles, and just about everything else we consciously experience. What cannot be accounted for by the application of these modes of thought and action? Learning itself. The process that brings these organizations into being out of chaos.

Control theory can help us define the goals of learning in an orderly way, and may be able to help us organize what is to be learned into a sequence that fits the sequence in which the hierarchy must be built up (you can't teach principles first —not if you want to teach what you intended to teach). But at least as important, it shows us what is being left out of the teaching process. Even without control theory, we know how to define what is to be learned—understanding, or perhaps performance objectives. We know how to set up a situation that requires learning to happen. But we do not know how to initiate, direct, or sustain that process itself. It happens, or it doesn't happen. Reorganization commences, or it does not commence. No teacher was ever told how to make it start.

This process of change is not an epistemological problem. It is not concerned with the product of learning, which is the perceptual hierarchy that does or does not represent a world outside of perception, and which is part of a control-system organization. Because of that fact, an approach to the subject of reorganization, and especially to its relationship to human needs and built-in human properties even deeper than what we call needs, may offer a totally new approach to epistemology. Perhaps the question we ought to be asking is: by what criteria does an organism decide that one representation of reality is "better" than another? We can easily say that a taste tastes "good." We cannot so easily ask whether "good" is good. This brings us close to questions about our very nature, and its relationship to all living things.

Maybe this is the most suggestive result of my epistemological investigations: recognition of the fact that the process of change is not part of the epistemological problem, but creates the kind of organization that puzzles about epistemology.

Reference

Powers, W.T. *Behavior: The Control of Perception.* Chicago: Aldine Publishing Co., 1973.

[1976]
The Cybernetic Revolution in Psychology

The picture of a cybernetic model of an organism which I will present in this essay represents what has condensed out of an amorphous cloud which has floated over the United States, Italy, Russia, and England, continuously changing shape but always seeming to gather itself into a more and more definite form. While much is still indefinite, the fundamental principles of a new concept of human and animal nature are now clear. They have nothing to do with automation, man-machine relationships, the study of vast social systems, or the creation of a cybernetically planned political system; some cyberneticists will be as disappointed by that as some psychologists will be relieved. Instead, these principles seem to point in the direction of individual autonomy and freedom, and a level of individual responsibility some might find daunting. I will not pursue such conclusions, however. The main aim here is to present, as clearly as possible, a set of ideas which are likely to cause some of the most fundamental assumptions of behavioral science to be discarded.

Background

Cybernetics began when Norbert Wiener[1] and his associates saw the parallels between the organization of automatic control systems and certain neuromuscular organizations in living systems. This occurred in the middle of a technological

Copyright 1976 by the American Society for Cybernetics. Reprinted with permission from *ASC Cybernetics Forum* 8(3 & 4), Fall/Winter 1976, 72-86.

explosion and a war, so perhaps it was inevitable that the accent lay on the technology rather than the living systems.[2] Wiener himself became very concerned with the impact of automation on society, and spent the rest of his life trying to warn us of the consequences of mishandling that technology.[3] Other cyberneticists followed other trails,[4] but most of them seemed bent on reducing human behavior to a technological model of one sort or another. Many of those trails, I believe, were false. The search for general theorems concerning the properties of social and man-machine systems, the efforts put into applications of information theory and digital computer models,[5] and the attempts to find sequential-state neural models that would reproduce behavior[6] were all, I believe, off the main track onto which Wiener, knowingly or unknowingly, set us. Some admirable talent went into these efforts, but despite them cybernetics has not yet lived up to its original promise of providing a dramatic new understanding of human nature.

There are several reasons for that temporary failure. In my opinion, the primary reason was that the leaders of the cybernetic movement either were or tended to become involved in a search for the most general possible mathematical theorems, general enough so that any conceivable human action or interaction could be treated as a special case. As a result, the pursuit of complexity postponed the understanding of *simple* relationships.[7] Not many who led that movement had ever designed and built a control system, or cursed and sweated to make it work properly, or experienced any extended personal interactions with a working control system; the interactions tended far more to be between cyberneticist and block diagram.

There is a world of education in the simple experience of pushing on a control system and feeling its firm and instantaneous resistance to the push; it literally feels alive. It feels the way a human organism feels when you push on it. It pushes back. The explanation for this behavior is not that human organisms are really nothing but soft technological machines, but that servomechanisms were modelled after human behavior in the first place. The engineers who were designing devices to take over tasks formerly requiring a human operator would have been insulted to be told they were psychologists, but as they studied what a person had to do in order to carry out a control task, and boiled those requirements down to the basic working components of a control system, they were modelling human organization just as certainly as if they had intended to do so.

The concepts behind control theory, developed during the 1930s, could not have been discovered by the psychology of the 1930s, because scientific psychology was then convinced that no physical system could have the properties a control system has. The phenomena to be seen in control behavior involve subjective perception, goal-selection, and intentions; psychology had been engaged for some thirty years in reaching a consensus that such notions were metaphysical and had no place in science. Thus, American and other psychologists were busy explaining away the very properties of life which control engineers, living in a different universe, were discovering to be essential in any control organization. That is why control theory had to enter through the back door, out of electronics, through Wiener, and thenceforth via a few mavericks who could call themselves less psychologist than engineer. That is why I am writing here about the earliest and simplest of the cybernetic notions, the notion that organisms act as control systems. It is time this idea got into psychology without being so distorted as to preserve basic assumptions which are really totally incompatible with it.

In order to say anything about the cybernetic analysis of human behavior, it is necessary to have a model of what a human being *is*. I say that human beings are self-reorganizing hierarchies of negative feedback control systems. Now that we have seen how little a string of words can mean, I hope it is obvious that before getting to the subject I have to begin by talking about what a control system is. Before I can talk about what it is I have to define what it does.

What a Control System Does

A control system controls some physical variable, or some function of a set of physical variables, *outside itself*. That statement leads another step into detail: what does "control" mean in the context of interest here?

The central concept that has to be defined is *controlled quantity* or *controlled variable*. I prefer "quantity" to remind us that control is always quantitative. What is there about a quantity that distinguishes it from other quantities and makes it become a controlled quantity?

The distinction can only be made in terms of a behaving system that senses and affects that quantity. The sensing part comes from control theory, which shows that no system can control anything except what it can sense; we do not need that part of control theory however, to identify controlled quantities in a clear way that has experimental significance.

Let us talk about controlled quantities, actions, and disturbances. An action is something a behaving system—an organism—does to its surroundings, in a way that depends entirely on the organism's activities and is not subject to direct external interference. A disturbance is a physical quantity in the environment that can be caused to take on any value, without regard to the action. A controlled quantity then belongs to the class of all quantities that are *jointly* affected by the action and by the disturbance.

Generally when the effects of an action and of disturbances converge on some physical quantity in the environment, that physical quantity will be caused to change in some way, but there will be no regular kind of change, save accidentally, since action and disturbance have no common cause. A controlled quantity, however, is an exception. We say that a jointly affected quantity is a controlled quantity if and only if for every magnitude and direction of effect a disturbance has on it, an action has an equal and opposite effect, with the result that the controlled quantity does not change.

This formal definition can be translated into laboratory methods for discovering human control systems and then evaluating their properties; I will get into that later. But it can be put much more simply. A quantity is controlled if, when one attempts to alter its state by pushing on it or otherwise performing acts that should physically influence it, something else pushes back and keeps it from changing. If one is satisfied that the quantity would have changed as a result of the push if only the pushing-back had not occurred, a control system has been discovered even if one does not yet know where it is. Any behaving system that can create this result consistently (when trivial explanations are ruled out) is a control system; that is what we mean by a control system, and why we have built so many of them. Control systems are capable of stabilizing aspects of their surroundings against disturbances of any kind, predictable or unpredictable, familiar or novel.

Control phenomena are difficult to explain in terms of any traditional cause-effect model. Suppose one controlled quantity proved to be the amount of food delivered into a foodtray and ingested daily by a rat. This would mean that any change in the environment of the rat tending to alter the amount of food delivered and ingested would be countered by an alteration in the actions of the rat which kept the amount of food delivered and ingested, per day, from changing. If we believe, as most scientific psychologists have believed, that behavior is caused by stimuli impinging on the sense-organs of animals, we must somehow find a stimulus which not only affects the sense-organs to create very precisely defined and quantitatively determined changes in actions, but continues to do so despite continuous alterations in the rat's position, and produces just the change of behavior at all times that will cancel the effects of the disturbance. If the disturbance changes, the stimuli have to change and affect behavior in just the way that will leave the joint consequence of action and disturbance, the daily food intake, undisturbed.

It would be easy to explain this strange matching of behavior to disturbance, and the stranger constancy of the consequence, if we were allowed to say that the rat wants to eat a certain amount every day, and simply does whatever it has to do in order to get that amount of food. That's the simple-minded, common-sense explanation, but it was specifically rejected by scientific psychology. In scientific psychology, there is no way to translate terms like "want" or "in order to" into terms of the deterministic cause-effect model that has always been accepted.

As a result scientific psychologists have had to find *other* explanations, or else, like B.F. Skinner, give up finding explanations altogether and claim just to record the facts.[8] Those other explanations have become extremely involved creating a cancerous growth of jargon as a result and little agreement among factions. It is very difficult to handle control phemomena without using the common-sense model, for the simple reason that the common-sense model has been the right one all along, and the others based on simple determinism are wrong. It takes a great many words to make a wrong model seem to be the right one.

A somewhat more precise way to state the common-sense model is to say that organisms control what they sense relative to internally specified reference levels. Controlled quantities and their relationship to actions and disturbances are the externally observable parts of a control process, the rest of it happening inside the system doing the controlling. We can thank those non-psychologists, the control engineers of the 1930s, for having discovered the kind of internal organization a system must have in order to create these external appearances.

How a Control System Works

It is astonishing how little difference there is between a control-system model of an organism and the old stimulus-response model. Only two relatively minor modifications have to be made, for a stimulus-response organism to become a control system.

First, the concept of a stimulus has to be broadened to include something more than instantaneous events. Most sensory stimuli, after all, are continuous variables that always have *some* magnitude, and while sensory nerves show some hypersensitivity to rate of change, they are far from insensitive to steady values of stimulation.[9] So we must think of stimuli as continuous variables, not just pokes and jabs. Of course responses, too, have to be generalized to include continuous outputs.

The other change seems even more innocuous. Instead of assuming that a behavioral output goes to zero only when the stimulus input goes to zero, we will treat that as a special case of a more general input-output relationship: the output action is a function of the input stimulus, but goes to zero when that stimulus is at some particular value we can call the *reference level*. In many cases this has no more effect than re-defining the zero-point of the physical scale of measurement of the stimulus. The old model says in effect that an organism will do nothing if it is not stimulated to act. The control model says that the organism will do nothing if the stimulus input matches some particular reference level of input. That may hardly seem worth writing a paper about, much less calling it the basis for a revolution, but we shall see.

Cybernetics has nothing on psychology when it comes to recognizing feedback phenomena. John Dewey, in 1896, was preaching (to deaf ears, unfortunately) that the reflex arc could not be divided into causes and effects because it was a closed loop.[10] Many other psychologists, especially Thorndike in the early days,[11] recognized that behavior has effects on the organism itself—on what it is sensing and on its physiological state. In fact these ideas led to several stimulus-reduction or drive-reduction theories, which logically concluded that behavior tended to reduce stimulation, and hence that organisms always tended toward the state of total lack of stimulation.

That quite logical conclusion didn't hold up, because organisms do not always seek zero stimulation. A lizard moving from the shade into the sun refutes stimulus-reduction theory. But that theory is really a control-system theory; its only fault is that it assumes reference levels always to be set to zero.

When we recognize that a reference level can have *any* value, zero being only one example, we can immediately generalize from stimulus-reduction theory to stimulus *control* theory; i.e., to control-system theory itself. Organisms tend to bring stimulation not to zero but to specific reference levels. How they do this, once the basic requirement is understood, is not hard to see.

The reference level represents some kind of bias in the perceptual process. Let us start with that bias set to zero, and see how action and disturbance would relate to a controlled quantity that was also a stimulus input to the organism. If we start with the stimulus input at zero, there will be no action. Now we introduce a very small constant disturbance acting on the controlled quantity to raise the level of stimulus input slightly above zero. This small input causes some amount of output, the exact amount depending on the sensitivity of the organism to small steady stimuli. The continuous output, the action, affects the controlled quantity and hence the stimulus, and the feedback loop is closed.

If the effect is to *further increase* the stimulus, the feedback is positive, and if the organism is at all sensitive this will lead instantly to a runaway condition; the small increase of stimulus due to the action *aids* the effect of the disturbance, making the net stimulus larger and leading to a still greater effect of action on the stimulus. Only if the organism has a very low sensitivity to stimuli will this system not run violently to its limit of output. There are few examples of positive feedback in behavior.

If the feedback is negative, the output will be designed to affect the stimulus in the direction that makes it *smaller*. If the organism is highly sensitive to small steady stimuli, only a very small amount of stimulus will be needed to produce enough action to cancel most of the effect of the disturbance. As the disturbance is made larger and larger, the stimulus grows and grows, but since it therefore produces more and more action opposing the disturbance, it does not grow very much. If the organism were highly sensitive to stimuli (and a few other design details were properly taken care of), this system would effectively cancel the influence of any disturbance on the stimulus input; that input would be actively held at zero. The action would continuously and precisely balance out the effects of the disturbance. In real control systems this condition of exact balance is very closely approximated, and all but the poorest control systems can be treated for most purposes as if they were ideal.

Now let us change the bias on the perceptual apparatus, which is to say we will set the reference level at a non-zero value. A non-zero level of input will *look* to the system like zero input.

We begin as before with the stimulus input at the reference level: that now means there is some finite level of stimulation, but the output is zero because the bias just cancels that amount of stimulation. Introducing a steady disturbance that tends to *increase* the stimulation, we find exactly the same situation as before except that the effective stimulus is now not the actual stimulus magnitude, but the excess over the reference level. A steady disturbance tending to increase the amount of stimulus above the reference level will, as before (for negative feedback), result in a steady degree of action that cancels most of the effect of the disturbance; the stimulus rises slightly above the reference level, but the more sensitive the organism is to what we will now call error, the less will be the change in the stimulus needed to bring about a properly opposing output.

Previously we could consider only one sign of disturbance and of output action, since there could be neither less than zero stimulation nor less than zero output. Now that the reference level is moved away from zero, we can have disturbances tending to *decrease* the stimulation. A suitably constructed system would then produce the opposite sign of output, as the stimulus fell *below* the reference level (because of a disturbance or spontaneously) and the opposite sign of error was created. As before, if the system were sensitive enough, only a very small fall would generate all the output needed to cancel most of the effect of the disturbance.

Now we have a model in which a stimulus input is actively held at a particular reference level, different from zero, despite disturbances that tend *either to increase or to decrease it*. We have, in fact, a complete control system that will behave exactly in the way needed to satisfy the definition of external controlled quantities. The stimulus, or whatever physical measure we use to define it, will prove to be a continuously controlled quantity.

Reference Levels and Intentions

The reference level of the stimulus input is determined by the bias that exists in the perceptual processes inside the organism. It is *not* determined by the stimulus but by the organism. A given amount of the controlled quantity might lead to behavior that tends to increase the controlled quantity, or to decrease it, or to do nothing to it; which will happen depends entirely on the amount of perceptual bias, the setting of the reference level.

The reference level in a given control situation is a property of the organism, not of the environment; it is the organism's perceptions that are biased, not the physical variables outside it. That bias determines whether a given stimulus input will be treated as *not enough, too much,* or *just right.* Therefore this perceptual bias or reference level specifies the amount of stimulus input which the behavior of the organism will create and maintain despite all normal disturbances.

A strong, fast and highly error-sensitive control system does not "correct errors" or "seek goals." Being designed to work properly in an environment where some maximum amount of disturbance is possible, and where disturbances are expected to vary in magnitude no faster than some maximum speed, a good control system never permits the controlled quantity to stray significantly from its reference level. Its actions always provide whatever forces are needed to maintain the controlled quantity at the reference level, and whatever further increases or decreases of forces are needed to oppose the effects of disturbances. The only time we see a good control system correcting an error or seeking a goal is just after it has been turned on, or just after the termination of some disturbance larger and faster than it can cope with.

An accurate way to describe an organism acting as a good control system is to say that it *carries out intentions.* This common-sense term, intention, reflects an intuitive understanding of reference levels, although a confused understanding of what they pertain to. Whatever is being controlled, there is an intended state relative to which that control takes place. We have seen that this state is determined by a perceptual bias that defines the "just-right" state; hence what is intended is not an action or an objective external consequence of actions, as common usage suggests, but the state of a perception.

Perception and Control

In the course of simply living our lives, we human beings do not constantly refer to philosophical reflections about the nature of reality: there is a world of appearances and we learn to deal with it. When we speak of an intention, such as the intention to go to Kansas City, we imagine some future state of the external world, some objective state of affairs with a little picture of ourselves in it. In imagination we can take any point of view we like, and call it past, present, or future.

When it comes time to carry out that intention, however, we must always work in terms of present-time perceptions, and even if we continue to project those perceptions into a real objective world, all we can actually control are the perceptions: we will or we will not perceive ourselves to be in the place we perceive as Kansas City. If there are errors—if we get the uneasy feeling that this looks like St. Louis—we can only base our actions on the difference between what we *are* perceiving and what we *intend* to be perceiving. It makes no difference whether our perceptions are exact copies of reality or transformations of a reality with utterly different dimensions; it is always the perception, not the reality, we must control, here and now.

All control, artificial or natural, is organized around a representation of the external state of affairs. If that representation is created in a consistent and quantitatively stable way, controlling the representation to keep it in a particular state will, presumably, entail actions that bring the external state of affairs to some corresponding state. As it were, we reach around behind our perceptions and manipulate whatever it is that is causing them until the perceptions look just right, in precisely the way one reaches behind an alarm clock and adjusts the alarm pointer until he sees it in the right position. If physics has taught us anything in the past three hundred years, it ought to be that we do *not* experience the actual effects of our actions in the physical world; we experience only their perceptual consequences, those capable of being represented in human nervous systems. Most of what goes on out there connecting action to perception is not experienced at all. When we reach around behind the alarm clock we *feel* the knob and the turning efforts, and we *see* the pointer move, but most of us have no accurate knowledge of how the feeling causes the seeing, inside the clock's case.

The object of behavior, therefore, is always the control of perceptions relative to reference levels.

Variable Reference Levels

Good control systems never allow perceptions to stray far from their reference levels: if that is the case, how can there ever be *dynamic* controlled perceptions? There are really two answers to this question; we will postpone the answer having to do with types of perceptions.

One important way in which controlled perceptions can be caused to change is for the reference level to change. The biaspoint of the perceptual system determines the level of the perception that is just right—that calls for no action—and thereby defines the state toward which all actions will force the perception via effects in the external world. If that bias changes, what was formerly the "just-right" condition of the perception will become "too little" or "too much," depending on whether the bias increased or decreased. If the perception is now "too little," the negative error will result in actions that increase the level of the perception; if "too much," the positive error will drive actions that decrease the level of the perception. In a good control system only a small error is required to produce a large change in output, and it doesn't matter what caused the error—an external disturbance or an internal change in reference level. The perception will be kept at the reference level, and now we see that this means the perception will *track* a *changing* reference level.

We can begin to see how a hierarchical control model is constructed. If we associate one subsystem with each kind of controlled quantity, we can see that anything capable of varying the perceptual bias of a subsystem will vary the reference level for the controlled quantity of that subsystem. The subsystem will cause the controlled quantity to track the varying reference level, so in effect whatever can vary the perceptual bias can alter the state of the controlled quantity, the control subsystem providing the action and taking care of disturbances all by itself. The command that alters the perceptual bias does not tell the subsystem what actions to perform; rather, it tells the subsystem how much of its perception it is to create. The action will correlate with the command only in a constant environment. When varying disturbances are present, the major part of the action will be directed so as to oppose the errors which the disturbances would otherwise cause, and there may well be no discernible correlation between command and action. The command, in effect, specifies a perception, not an action. Higher centers in the brain do not command lower systems to create certain behavior patterns, but certain patterns of perception.

Types of Controlled Quantities

In order for an aspect of the environment to become a controlled quantity, it is necessary for a control system to exist which can sense the current state of that quantity and by taking action affect that state. The causes of changes in the controlled quantity, extraneous disturbances, do not have to be sensed directly; only their effects have to be sensed, and those effects are already taken care of by the fact that the control system operates on the basis of error, and acts directly on the controlled quantity to oppose error.

The phrase "aspect of the environment" does not, as one might at first suppose, refer to some objective property of the external world; it refers to perceptual processes. In any given environment, there is an infinity of different quantities that might be controlled. What they are depends, of course, on the raw material from which perceptions are constructed, stimulation from outside, but it depends even more crucially on how a given perceptual system combines the lowest-level sensory signals into higher-order variables.

There is no need to think of all controlled quantities as simple physical variables: force, angle, position. Human beings are equipped to perceive not only such elementary variables, but highly complex functions of such variables.[12] For instance, one can perceive not only the position of a passing automobile, but the rate of change of position, which we name speed. One can perceive, at every moment, the speed of a pendulum of a clock, but perceivable at the same time are higher-order variables that are functions of speed and position: the amplitude and period of the swing. One can perceive events or finite sequences such as a tennis serve, a backhand volley, or a lob, but these become variables in a higher-level function perceived as a *relationship* among independent sets of such elements: two persons controlling such finite sequences *in relationship* are playing tennis.

The most interesting human relationships are those in which each person, in controlling his own perceptions, disturbs the perceptions being controlled by others.

The general pattern of hierarchical control is this: in order to control perception of one level, one must control at least some of the lower-level perceptions of which the higher is a function. The lower-level perceptions which are not under control by lower-order systems are disturbances; as the uncontrolled perceptions vary, the controlled ones must be adjusted so that the net result at the next higher level matches the reference level.

Clearly, a constant state of a perception at many of these higher levels entails a constantly varying state of lower-level perceptions; a driver's constant impression of the speed of his automobile entails a constantly-changing configuration of the visual field. A person cranking a bucket out of a well perceives and controls a constant cranking speed or angular velocity, maintained by continuous variations of arm position. This illustrates the other way in which control of a perception relative to a fixed reference level can lead to dynamic conditions.

The object here is not to develop any specific hierarchical model; that is too much to cover in a short essay. It is primarily to introduce the kinds of relationships that exist between levels of control, and even more to the point simply to broaden the concept of what a controlled quantity can be. Anything that a person can sense and affect, regardless of its nature or its objective existence, can become a controlled quantity for that person.

The Meaning of Empirical Correlations

The phenomena of control in human behavior are organized around a highly subjective set of perceptions. In order to understand what a person is doing, it is necessary to understand what that individual is perceiving. Simply watching a per-

The Cybernetic Revolution in Psychology 117

son's actions is insufficient; all that will do is show, indirectly, the presence of disturbances. It will not reveal *why* certain environmental events are accompanied by certain actions. To understand why there is a relationship between events and actions, it is necessary to find the controlled quantity being affected both by the independent event and by the action, and to understand how control systems work.

Under the old paradigm, it is impossible to discover controlled quantities. The traditional approach to an analysis of behavior is to select some set of environmental variables that could determine a given behavior, and to vary them while looking for correlations with changes in the behavior. When a high correlation is found, it is assumed that the stimulus or situational variable is acting on the organism to make it produce a corresponding change in behavior. The organism is thought of as *mediating* between cause and effect; between the manipulated variable and the resultant change in behavior. If changes in behavior are observed to be a regular function of the manipulated variable, it is assumed that the form of that function describes the organism's transfer function, the overall process in the organism between its input and its output.[13] What is normally called a stimulus or a "track input" is what we are calling here a disturbance.

Control theory shows that such a transfer function is an illusion, in any case where a controlled quantity exists. When a controlled quantity is being stabilized against disturbances, the observable relationship between a disturbance (stimulus) and the system's action is dictated completely by the physical connections, external to the behaving system, from disturbance to controlled quantity and from action to controlled quantity. Whatever effect the disturbance tends to have via its physical link to the controlled quantity, if the controlled quantity remains undisturbed, the action must be precisely the one which, acting through the physical link from action to controlled quantity, continuously cancels the effects of the disturbance. The more nearly ideal the control system, the more exactly will the relationship of action to disturbance be predictable strictly from an examination of these physical relationships in the environment. The transfer function that is observed describes external physical relationships, and very little about the organism.

If the controlled quantity is not of an obvious nature, and few of the interesting ones are, the link from either action or disturbance to that controlled quantity may be subtle and indirect. Thus not every way of measuring action or disturbance will be appropriate.

If I am controlling the position of a car on the road, forces caused by the muscle tensions I use to turn the steering wheel may not often be exactly tangential to the wheel; a direct measure of muscle tension as a measure of my behavior, therefore, will include force components that are unrelated to the control task, and which will introduce noise into the observed relationship. Furthermore, the car's position is not affected directly by muscle tension, but by the effects of those tensions transformed by intervening mechanical linkages and by laws of mechanics involving at least two time integrals. A direct measure of muscle forces might show some significant correlation with a direct measure of the car's position, but it would not be a very high correlation.

For the same reason, the effect of a crosswind on the same variable will be indirect and complex, involving laws of aerodynamics and again laws of mechanics.

The tensions in my arm muscles will show some reasonably high correlation with wind velocity, and it may seem that in my soundproof car with the windows rolled up I am still able to sense and respond to wind velocity, but neither the correlation nor the apparent response to the stimulus is real. If one were to calculate the effect of muscle tension on car position using the correct physical principles, and also compute the effect of wind velocity on car position using the correct physical principles, it would be seen that the relationship between the properly transformed measures would be *exact*: the effects, integrated, would be equal and opposite. It would not be merely a statistical correlation, but a continuous and precise quantitative relationship. And it would be obvious that this exact relationship reveals almost nothing about the behaving organism.

The presence of a control system creates an apparent cause-effect relationship between disturbances and actions. This relationship is what scientists have been studying since the beginning of the life sciences. It tells us very little about the organization of the behaving system, except the bare fact that it is acting as a control system. That is, of course, an important fact to know, but it does not need to be proven ten thousand times.

Discovering Controlled Quantities

The behavioral scientist is not in the same position as a servomechanism engineer who works with a known system controlling known variables. The behavioral scientist sees only the multitudinous actions of an organism and variations of immense multiplicity in its environment. He does not know in advance which effects of motor behavior are parts of control actions and which are merely side-effects; he does not know which extraneous events are significant and which can be ignored. If he is studying a human being, any aspect of the environment that the experimenter might notice could prove to be a quantity under control by the behaving system. It is not very likely, if the experimenter simply attends to aspects of the environment or of the other person's behavior that are interesting to himself, that he will happen across variables that are actually of any behavioral significance. Some sort of systematic procedure is required.

As a start toward discovering controlled quantities, one can look for regular relationships between disturbances and actions. In this regard, previous empirical searches for behavioral laws will not have been entirely wasted, although they will almost certainly have stopped short of revealing the final object of such a search. When a regularity linking extraneous events and behavioral actions is found, this is a hint that there may be a controlled quantity being jointly affected by both. That quantity, if controlled, will be hidden precisely because it does *not* alter in response to changes in the disturbance or in the action. In order to find the controlled quantity, one must understand the physical situation well enough to detect variables that do not change when they ought to change.

Most physical quantities or aspects of the environment that are jointly affected by an action and a disturbance will show variations which reflect the resultant of both effects. Since most such jointly affected variables will *not* be controlled quantities, their variations *will* show significant correlations both with disturbance and with action. Examination of the details will reveal that the variations are simply those due to variations in two independent causes, both of which can affect the quantity but which are uncorrelated with each other.

A controlled quantity will be identified when a variable is found that is affected equally and oppositely by action and disturbance. The action itself does not have to be "equal and opposite" to the disturbance itself, nor will it generally have any effect on the disturbance. It is the *effect of each one on the controlled quantity* that must be equal and opposite to the *effect* of the other. If the measuring instrument were affected by the disturbance and the action in exactly the way the controlled quantity is affected, the equal-and-opposite relationship would be obvious, but most convenient measuring instruments will not be affected that way. A device for measuring muscle tension in a driver's arms will be calibrated in dynes or pounds, not in units of change in car position. A device for measuring wind velocity will be calibrated in units of dynamic pressure, not in feet of displacement relative to the center of the road. Before one can evaluate either action or disturbance, therefore, it is necessary to apply the correct transformations to the direct measurements, and one cannot know what transformations to apply until the controlled quantity has been identified.

The Test for the Controlled Quantity

Therefore, controlled quantities cannot be found by deduction, but only by induction. One must make an intelligent guess as to the nature of a controlled quantity, and then test that guess.

The test for the controlled quantity is carried out as follows. Given a definition of a controlled quantity to serve as the hypothesis, one searches for physical links from action to the controlled quantity and from disturbance to controlled quantity. Those links are analyzed in terms of physical laws, and the effects of action and disturbance are separately calculated in units of effect on the controlled quantity. The predicted sum of those effects must be zero (or zero change), and the measure of the controlled quantity must correspond to that prediction, if the hypothesis is to be accepted.

The experimenter does not generally have the ability to predetermine the organism's action, but he is free to select and vary disturbances at will. Thus one way to apply the test is to select disturbances which, through known physical relationships, are capable of altering the state of the proposed controlled quantity when acting alone. If the *observed* variation in the quantity is only a small fraction of the calculated change, that quantity is likely to be under control. To complete the proof that an organism is in control, one must then show that the reason for failure of the controlled quantity to change is that the action of the organism, working through known physical links, is continuously cancelling the effects of the disturbance that were calculated. Further support of the hypothesis requires showing that the controlled quantity *must* be sensed by the organism in order to be controlled.

More conventionally, the hypothesis is *disproven* if applying a disturbance to the proposed controlled quantity succeeds in disturbing it as if only the disturbance were acting. There is a gray area that calls for judgements; if the effect of the disturbance is less than predicted, but not dramatically less, the chances are that the defined quantity is related to a controlled quantity but not identical to it. One must then select some criterion for "good enough" proof—proof that will permit one to proceed on the assumption that the quantity is a controlled quantity. One might decide, for example, that if the observed effect of a disturbance is more than 10 per cent of the predicted effect, the quantity is not controlled. The criterion level will depend on how well one expects the organism to be able to control variables of that kind, even when they are defined perfectly correctly. Most human control systems can cancel 90 per cent of those disturbances lying within their range of control; not a few can cancel 95 per cent or even 99 per cent. Controlled quantities discovered by this method are normally clear-cut; there is little need to consider "statistical significance," although occasionally that is appropriate.

The Test for the Controlled Quantity is a direct nonverbal

experimental procedure that teaches the experimenter to perceive the environment in essentially the same terms in which the behaving system is perceiving it. Neither the experimenter nor the control system needs to know the actual physical situation underlying the controlled quantity: in other words, epistemological questions are bypassed by the Test. I believe the Test to be the first scientific method by which an experimenter can come to know the subjective world of his subject without involving the medium of symbolic communication. It will work with anything that behaves.

Cybernetics and Behavioral Science

The cybernetic model of a behaving organism is fundamentally different from the model which has been assumed for over three hundred years, in all branches of biology, physiology, neurology, psychology, and the social sciences. Even those schools of thought which profess to abhor mechanistic explanations revert to the old cause-effect model when it comes to testing hypotheses in the framework of scientific method: they still manipulate condition A and look for correlated changes in behavior B. To many psychologists, this is simply scientific method itself; whatever hypothesis one may make concerning inner processes, one must finally put those hypotheses to the test in a cause-effect setting.

The cybernetic model is based on a new principle of organization in which closed-loop relationships exist. Before control theory was fully developed, no person on earth understood how such systems could exist, or why their mode of operation could not be described in the simple cause-effect terms that apply to inanimate systems. Without control theory to point out the possibility of controlled quantities, no scientific interpretation other than a simple deterministic cause-effect one was possible, for the appearance is that disturbances are stimuli that act on organisms to make them behave.

In the light of control theory we can now understand some of the most baffling phenomena that have been noticed, particularly the peculiar *rationality* of behavior. Under the old deterministic picture, it was impossible to explain why stimuli should so kindly combine their effects to produce just the behavior that was good for an organism; a great deal of qualitative arm-waving has taken place in the attempt to give the impression of an explanation, but such explanations have had little explanatory power and no predictive power at all. If a rat pressed a food-delivering bar often enough to keep its own body weight constant, that result was normally treated as a piece of good luck for the rat.

When B.F. Skinner concluded that behavior is controlled by its consequences, he came the closest of any psychologist to discovering control theory in its original context, that of behavior. But this statement taken literally is an affront to determinism, and belongs with certain other concepts such as "retroactive inhibition" that are phrased as a challenge to know-it-all physicists. Consequences are, as far as anyone knows, caused by their antecedents, not the other way around. Behavior in a Skinner box is not caused by the food it delivers to the animal; quite the contrary, the rate of food delivery is determined by the behavior, via the properties of the apparatus. No behavior, no food. There is no need to state the obvious situation any other way.

The statement that behavior is caused by its consequences can be converted easily into a correct statement of how a control system works if we add just one phrase at the end: "... relative to the consequence the animal wants." The organism always acts to keep the consequences of its behavior, as they affect the organism, matching the reference-consequence determined inside the organism, not in its environment. As in the case of early notions of drive reduction or stimulus reduction, Skinner's formulation missed the key concept by omitting the concept of the reference level.

It was not possible for Skinner or any earlier approximators of control theory to follow through to the correct conclusion, because the proper train of thought was cut short by an assumption so strong as to be impervious to reason: the assumption that physical determinism required all behavior to be caused from outside the organism.

Control theory shows that assumption to be false; the principal determinants of behavior lie inside the organism, and ultimately trace back to the inherited requirements of survival, not to any present-time external events. The environment provides the setting in which the organism must achieve its fundamental requirements; the environment determines the links from action back to perception and from action back to physiological effects on the organism. But none of those reflected effects of action would imply the necessity for any particular behavior if it were not for the fact that inside the organism there are specified quantitative reference levels for those reflected effects. The organism requires that certain effects occur to a certain degree; it learns what to do to the external world in order to assure that they do occur (or that effects which must not occur do not occur).

A Cybernetic Model of Evolution

We cannot yet be sure which reference levels are acquired and which are inherited; essentially no work has been done on this question in the framework of control theory. But there must be some set of inherited reference levels, specifying conditions which must be sensed as holding true inside the organism, the specification remaining unaffected by the events of a single lifetime. These fundamental reference levels, not the nature of the external environment or particular events in that external environment, ultimately determine which consequences an individual organism will learn to create for itself. The actions which the organism performs will, of course, come to be those which, in the current environment, will in fact oppose disturbances while maintaining the required consequences at their required reference levels. This is the cybernetic picture of "adaptation." The organism adapts to the environment by altering what it does, but not by altering what it accomplishes by what it does, not in terms of the fundamental reference levels. To say that the organism adapts to its environment is to say that it alters what does not matter to it in order to maintain control of what does matter to it.

This concept of adaptation is even clearer in terms of a hierarchical model.[14] In such a model, lower-order reference levels become the means by which higher-order systems act; the higher systems freely adjust the lower-order reference levels, to create the lower-order elements of perception which will result in the specified higher-order perception. This creates a hierarchy of adaptations. A given subsystem can learn new actions when a change in the environment renders the old actions ineffective for control, but it cannot change the consequences it wants those actions to produce; only a higher-level system can change the reference level. The highest-level system must be the one that says, thou shalt breathe, thou shalt perceive harmony, thou shalt not experience hurt and illness. In whatever specific nonverbal terms the message is cast, the effect of all the inherited reference levels is to say, thou shalt live. That is the only reference level that will propagate through the ages.

All sorts of proposals have been made to explain the processes of mutation that create variations of parts of the genetic message. Control theory provides a rationale for suggestions that these changes are directed, not simply induced by cosmic rays or background radiation. Once the fundamental principles of control are understood, it is not hard to apply them even in situations far from the subject of present-time behavior. For example, it is not hard, in principle, to see how there could exist a kind of master control system at the chromosome level of organization, one which is a part of the microstructure of every cell. Control systems do not have to be made of vacuum tubes or transistors or even neurones; there are great biochemical control systems in every organism, and evidence of control systems even on so tiny a scale as to control the permeability of individual cell membranes. There is nothing farfetched about imagining a control system which acts by setting into motion slow processes of change at the level of DNA, in response to errors of the most fundamental kind conceivable —and for that reason, a kind scarcely conceivable. If the rate of variation depends on the amount of error, it does not matter whether the variation be systematically appropriate to the error; at least some of the changes will be appropriate, and those that are not will not propagate.

It is thus possible to see evolution itself as a control process, the same control process in every living thing. Donald T. Campbell has called the kind of process proposed here as "blind variation and selective retention,"[15] recognizing that the organizing principle is to be found on the input side, not the output side. What results from a process of blind variation depends on what is retained, and what is retained is determined by criteria for retention which, when met, result in termination of the process of blind variation. The most widely accepted concept of evolutionary processes makes the retention criterion simply a binary variable: survival or extinction. It would be far easier to explain speciation and variations within species if room were made for specific stored criteria that did not require extinction to see if they had been met. Control theory, by providing the concept of a reference level, provides a place where genetically-transmissible criteria can actively specify what consequences of blind variation shall be retained (i.e., shall terminate the variations); all we have to imagine is that the error between actual and specified states of certain fundamental quantities drives the variations. This model expects that the variation rate will be low for some organisms in certain epochs, and high during others. The organism ideally adapted to its niche experiences no error of the kind proposed (if not defined) here, and its rate of blind variation of the genetic message is at the lower limit. An organism evolving rapidly is suffering extreme error; it is varying the details of its genetic blueprint relatively rapidly. This model explains all that the simpler model of evolution explains, and more besides.

This model no doubt raises shades of Lamarck, the transmission of inherited characteristics. But by accepting the idea of *blind* variation, it avoids that trap. Furthermore, the time-scale on which this master control system works has to be taken into account. It is an example, perhaps, of a "sampled control system," one which works only at intervals, in this case perhaps just for a brief period in each generation, prior to sexual maturity, but long enough after birth for the consequences of current organization to have their effects at the cellular level. Since the magnitudes of the variables in the control system are stored in the DNA, and change only slightly with each active period, we would have to look at hundreds or thousands of generations in rapid succession to see the dynamics of control —to see how the rate of variation corresponds to error, and how error corresponds to changes in the external situation. In effect that is what paleobiology is already doing, but without this model as an organizing concept. I think it would be a fascinating project to see whether the basic reference levels of life itself could be discovered by such studies. "Survival" is too crude (and too verbal) a reference level to explain all adaptations.

One need not extend control theory to speculations about evolution in order to apply it to behavior, but I think that doing so creates a rather grand and coherent picture of life that is satisfying to one's sense of order. We can see the principle of the control system as possibly extending from the very beginnings of life to its most detailed expressions in present time behavior. Indeed, we can begin to see the principle of control as possibly providing the principle that makes all organisms one. *Life* adapts by altering what does not matter to it in order to maintain control of what does matter to it. What does *not* matter to it, in the long run, evidently includes being plant or animal, being of small or large size, being of one species or another species, or behaving in one way rather than another. What does matter? Something, I imagine, rather basic.

The Cybernetic Revolution

The analysis of behavior in all fields of the life sciences has rested on the concept of a simple linear cause-effect chain with the organism in the middle. Control theory shows both why behavior presents that appearance and why that appearance is an illusion. The conceptual change demanded by control theory is thus fundamental; control theory applies not at the frontiers of behavioral research, but at the foundations.

References

1. Buckley, W., *Modern Systems Research for the Behavioral Scientist*. Chicago: Aldine (1968).

2. Wiener, N., *Cybernetics: Control and Communication in the Animal and the Machine*. New York: Wiley (1948).

3. Wiener, N., *The Human Use of Human Beings.* New York: Doubleday (1974).

4. Ashby, W.R., *Design for a Brain.* New York: Wiley (1952).

5. Feigenbaum, E.A. and Feldman, J. (Eds), *Computers and Thought.* New York: McGraw-Hill (1963).

6. Arbib, M.A., *Brains, Machines, and Mathematics.* New York: McGraw-Hill (1964).

7. Von Foerster, H., White, J.D., Peterson, L.J. and Russell, J.K. (Eds), *Purposive Systems.* New York: Spartan Books (1968).

8. Skinner, B.F., Behaviorism at Fifty; *Science* 140, 951-958 (1968).

9. Granit, R., *Receptors and Sensory Perception.* New Haven: Yale University Press (1955).

10. Dewey, J., The reflex arc concept in psychology; in *Readings in the History of Psychology.* Dennis, W. (Ed). New York: Appleton-Century-Crofts (1948) p. 32-41.

11. Thorndike, E.L., *The Psychology of Wants, Interests, and Attitudes.* New York: Appleton-Century-Crofts (1935).

12. Hayek, F.A., *The Sensory Order.* Chicago: Univ. of Chicago Press (1952).

13. Poulton, E.C., *Tracking Skill and Manual Performance.* New York: Academic Press (1974).

14. Pattee, H.H., *Hierarchy Theory.* New York: Braziller (1973).

15. Campbell, D.T., 'Downward Causation' in Hierarchically-Organized Biological Systems; in *Studies in the Philosophy of Biology.* Ayala, F.J. and Dobzhansky, T. (Eds). Berkeley: Univ. of California Press (1974).

[1978]

Quantitative Analysis of Purposive Systems: Some Spadework at the Foundations of Scientific Psychology

The revolution in psychology that cybernetics at one time seemed to promise has been delayed by four blunders: (a) dismissal of control theory as a mere machine analogy, (b) failure to describe control phenomena from the behaving system's point of view, (c) applying the general control system model with its signals and functions improperly identified, and (d) focusing on man-machine systems in which the "man" part is conventionally described. A general non-linear quasi-static analysis of relationships between an organism and its environment shows that the classical stimulus-response, stimulus-organism-response, or antecedent-consequent analyses of behavioral organization are special cases, a far more likely case being a control system type of relationship. Even for intermittent interactions, the control system equations lead to one simple characterization: Control systems control what they sense, opposing disturbances as they accomplish this end. A series of progressively more complex experimental demonstrations of principle illustrates both phenomena and methodology in a control system approach to the quantitative analysis of purposive systems, that is, systems in which the governing principle is *control of input*.

This article concerns four old conceptual errors, two mathematical tools (which in this context may be new), and a series of six quantitative experimental demonstrations of principle that begin with a simple engineering-psychology experiment and go well beyond the boundaries of that subdiscipline. My intent is to take a few steps toward a quantitative science of purposive systems.

Qualitative arguments on the subject of purpose have abounded. Skinner (1972) has expressed one extreme view:

Copyright 1978 by the American Psychological Association. Reprinted by permission of the publisher from *Psychological Review* 85(5), September 1978, 417-435.

Science... has simply discovered and used subtle forces which, acting upon a mechanism, give it the direction and apparent spontaneity which make it seem alive. (p. 3)

An extreme opposite view is expressed by Maslow (1971):

Self-actualizing individuals... already suitably gratified in their basic needs, are now motivated in other higher ways, to be called "metamotivations." (p. 299)

In the middle ground are many others who have tried to deal with inner purposes, for example, Kelley (1968), McDougall (1931), Rosenblueth, Wiener, and Bigelow (1968), Tolman (1932), and Von Foerster, White, Peterson, and Russell (1968). I have contributed some arguments as well (Powers, 1973; Powers, Clark, & McFarland, 1960a, 1960b). Obviously, none of these arguments, which are all qualitative, has succeeded in settling the issue of inner purposes.

In the 1940s, many of us thought that the missing quantitative point of view had been discovered. *Cybernetics: Control and Communication in the Animal and the Machine* (Wiener, 1948) seemed to contain the conceptual tools that might at last explain how "mental" causes could enter into "physical" effects. It seemed that a bridge might be built between inner experiences and outer appearances. A cybernetic revolution in psychology seemed just about to start. Now, in the late 1970s, it is still just about to start. Something happened to the original impetus of cybernetics, as a river entering the desert splits into a hundred wandering channels and sinks into the sand. I have some suggestions as to what went wrong.

Four Blunders

It is not so much honest labor on my part that puts my name to this critique as it is a series of blunders (qualitative mistakes) by others who could have done long ago what I am doing now. However unavoidable, these blunders have been directly responsible for the failure of cybernetics and related subjects to provide new directions for psychology.

Machine Analogy Blunder

In 1960, the president of the Society of Engineering Psychologists wrapped up the previous decade of cybernetics as follows:

> The servo-model, for example, about which there was so much written only a decade or two ago, now appears to be headed toward its proper position as a greatly oversimplified inadequate description of certain restricted aspects of man's behavior.... Whenever anyone uses the word *model*, I replace it with the word *analogy*. (Chapanis, 1961, p. 126)

This view is still held. There are and have been for some time scientists who think of control system models of behavioral organization as a mere analogy of human behavior to the behavior of a technological invention. A little digging underneath the engineering models suggests that this opinion is mistaken. Servomechanisms have always been designed to take over a kind of task that had previously been done by human beings and higher animals and by *no other kind of natural system*, that of controlling external variables (bringing them to predetermined states and actively maintaining them in those states against any normal kind of disturbance; Mayr, 1970). It was not until the 1930s, however, that there existed a sufficient variety of sensors or electronic signal-handling devices to permit simulation of the more abstract kinds of human control actions, for example, the adjustment of a meter reading to keep an indicated pH at a predetermeined setting. The control-engineers-to-be of the 1930s necessarily had to study what a human controller was doing in order to see just what had to be imitated. The functions of perception, comparison, and action had to be isolated and embodied in an automatic system, a quantitative working model of human organization of a type that psychology and biology had never been able to develop. Thus, the servomechanism has always been only an imitation of the real thing, a living organism, and the engineers who invented it first had to be, however unwittingly, psychologists. The analogy developed from man to machine—not the other way.

Objectification Blunder

The machine analogy blunder set the scene for missing the point of control theory, but the objectification blunder would have been enough by itself. In cybernetics, it arose quite naturally out of the fact that artificial control systems are designed for use by natural ones, that is, human beings.

The designer and user of an artificial control system are understandably interested in the output of the system and effects of that output on the world experienced by the user. Control systems, however, control input, not output. When the input is disturbed, the output varies to oppose incipient changes of the input and thus cancel most ot the effect of the disturbance. Thus, the only way to make such systems useful is to be sure that the input to the system depends strictly on the environmental effect that the user wants controlled and to protect the input from all other influences. If that environmental effect is an immediate consequence of output, the output will appear to be controlled as far as the user's purposes are concerned. Indeed, the controlled consequence of the actual output is likely to be called the output.

Natural systems cannot be organized around objective effects of their behavior in an external world; their behavior is not a show put on for the benefit of an observer or to fulfill an observer's purposes. A natural control system can be organized only around the effects that its actions (or independent events) have on its inputs (broadly defined), for its inputs contain all consequences of its actions that can conceivably matter to the control system.

This was Skinner's (1938) momentous discovery. He concluded that behavior is controlled by its consequences, unfortunately expressing the discovery from the observer's or user's point of view. From the behaving system's point of view, however, Skinner's discovery is better stated in the following way: Behavior exists only to control consequences that affect the organism. From the viewpoint of the behaving system, behavior itself, as output, is of no importance. To deal with behavior under any model strictly in terms of its objective appearance, therefore, is to miss the reason for its existence. Cybernetics and especially engineering psychology simply took over this erroneous point of view from behaviorism. This error is closely related to the next one.

Input Blunder

Wiener himself was accidentally a principal contributor to the input blunder. A diagram from Wiener's (1948) book on cybernetics (see my Figure 1 for an adaptation of Wiener's diagram) was taken directly from an engineering and users' viewpoint model. Examining Figure 1, the reader will see that there is an "input" coming in from the left, which joins a feedback arrow at a "subtractor," or more commonly, a "comparator." The "error" signal from the comparator actuates the rest of the system to produce an "output," from which the "feedback" path branches. This basic form has been repeated without change in the literature of psychology, neurology, biology, cybernetics, systems engineering, and engineering psychology from 1948 to the present. It is nearly always interpreted incorrectly.

When a person concerned with sensory processes sees the word *input*, it is natural to translate the term to mean *sensory input* or *stimulus*. But the arrow entering the subtractor is not a sensory input. It is a *reference input*, and the information reaching the subtractor or comparator by that path is by definition and function the *reference signal*. Engineers show refer-

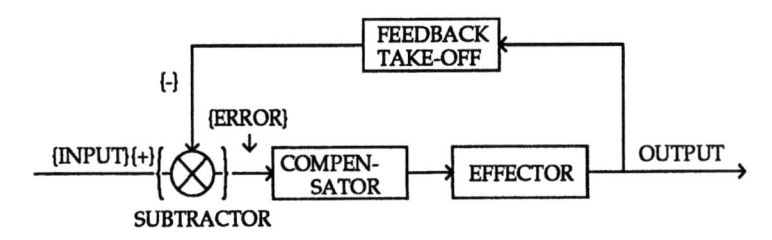

Figure 1. Adaptation of Wiener's (1948) control system diagram. (This diagram has misled a generation of life scientists. The "input" is really the reference signal, which in organisms is generated internally. Sensory inputs are actually at the input to the "feedback takeoff." Disturbances of the sensory input are not shown. Adapted with permission from *Cybernetics: Control and Communication in the Animal and the Machine* by Norbert Wiener. Copyright 1948 by M.I.T. Press.)

ence signals as inputs because artificial control systems are meant for use by human beings, who will operate the system by setting its reference input to indicate the desired value of the controlled variable. In natural control systems, there are no externally manipulable reference inputs. There are only sensory inputs. Reference signals for natural control systems are set by processes inside the organism and are not accessible from the outside. Another name for a natural reference signal is *purpose*. We observe such natural reference signals only indirectly as preferred states of the inputs to the system. Control systems are organized to keep their inputs (represented by the feedback signal) matching the reference signal.

Where, then, are the sensory inputs in Wiener's diagram? They are in the "feedback takeoff" position, or more precisely, they are in the junction where the feedback path leaves the output path. In that same junction are contained all the physical phenomena that lie between motor output and sensory input, which in some cases can include a lot of territory. The arrow labeled "output" and exiting toward the right should really be labeled "irrelevant side effects" because effects of output that do not enter into the operation of this system are of importance only to some external observer or user. Those side effects tell us nothing about the principles of control.

Man-Machine Blunder

If one's primary purpose is to keep pilots from flying airplanes into the ground or to make sure that a gunner hits a target with the shell, that is, if one's purposes concern objectivized side effects of control behavior, the man-machine blunder amounts to nothing worse than a few mislabellings having no practical consequences. If one's interest is in the properties of persons, however, the man-machine blunder pulls a red herring across the path of progress.

Consider Figure 2, adapted from Poulton (1974). The "man" in this experiment is supposed to hold a cursor on the display next to a fiduciary mark; this task is like keeping a ship on a compass course or flying an airplane level by keeping an artificial horizon centered. The immediate task is to maintain a given appearance of the display; a side effect of doing so is to stabi-

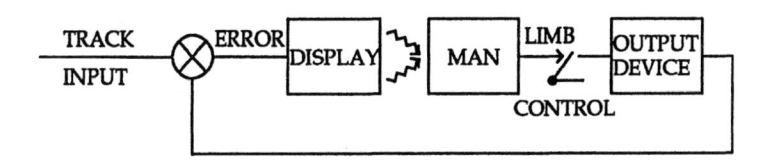

```
TRACK ──⊗── ERROR ──DISPLAY──┐ ┌──MAN──LIMB──┐ OUTPUT
INPUT                                          DEVICE
                              CONTROL
```

Figure 2. Compensatory tracking. (The "man" is a stimulus-response device embedded in an artificial control system. The influence of Wiener's [1948] diagram is apparent [see the present Figure 1]. Adapted with permission from *Tracking Skill and Manual Control* by E. Poulton. Copyright 1974 by Academic Press.)

lize some objective situation of which the display is a partial representation. The objective situation, of course, is the whole point of the experiment from the experimenter's point of view.

From the subject's point of view, however, the display simply shows a variable picture that the subject can maintain in any stable condition desired. The subject could keep the cursor a fixed distance off the fiduciary mark, as a pilot could keep the artificial horizon above the center mark while deliberately losing altitude, or as a navigator-helmsman could keep the compass reading several degrees east of the intended course in order to compensate for a remembered westward deviation of magnetic north from true north.

The so-called "error" in Figure 2 is not an error at all; the error corresponds to a sensory input, both for the subject and for the experimenter. The crossed circle is not a comparator, but only a place where external disturbances join feedback effects in determining the state of the display. Wiener's diagrams did not show disturbances.

Only the subject has a means of directly affecting the state of the display; hence, the display will be made to match the subject's inner reference. If doing this causes the experimenter to see an error (Figure 2 shows the experimenter's point of view), the only corrective action the experimenter can take is to halt the experiment and persuade the subject to reset his inner reference signal to produce a result the experimenter experiences as zero error.

That, of course, is what is done. By demonstration and instruction, the subject is shown where to set his internal reference; if the subject complies, the experiment proceeds. The analysis of the data can then be done under the assumption that there is no offset in the "man" box. Thus objectifying the error assures that the experiment will not reveal one of the most important properties of the subject: the ability to manipulate an inner reference signal. As this situation is usually analyzed, the man's purposive properties drop from view, and those of the experimenter are quietly incorporated into the so-called "objective" analysis.

From the General to the Specific

The preceding discussion suggests that the failure of control theory to create a cybernetic revolution in psychology may not have been the fault of control theory. I hope my implied criticisms have stayed on target because there is no reason to belittle what cyberneticists have done or what engineering psychologists have discovered. The blunders I have described are principally blunders of omission and misinterpretation that have unnecessarily but unavoidably limited the scope of these endeavors. I shall commit blunders of my own just like these, as will we all. That is the penallty for trying something new.

In trying to develop control theory as a tool for experimental psychology, I think it is important to avoid assuming that any example of behavior involves a control organization. I have been critical of some psychologists for adopting a language that periodically asserts a model by calling every action a "response," but I succumb to the same kind of temptation myself when trying to convey my own point of view. A basic analysis cannot be very convincing if its conclusions are plugged in where the premises are supposed to go, so in the following section, the treatment will begin in as general a form as possible.

Let us assume little more than the early behaviorists did, and in some respects, let us assume a great deal less. The organism will be treated as nothing more than a connection between one set of physical quantities in the environment (input quantities) and another set of physical quantities in the environment (output quantities). By leaving the form of the organism function general, however, we will allow for possibilities that were tacitly ruled out at the turn of the century, the most important one being the possibility of a secularly adjustable constant term in the system function. That term will ultimately turn into the observable evidence of an inner purpose, although I will not pursue that point vigorously here.

This approach will explicitly recognize the fact that the inputs to an organism are affected not only by extraneous events but possibly by the organism's own actions. By leaving the development general, we will be able to deal deductively with feedback effects, not asserting them but simply stating the observable conditions under which they necessarily appear and those under which they can be ignored. Thus, the classical mechanistic cause-effect model will become a subset of the present analysis.

Let us now turn to mathematical tools, beginning with an approach that is neither as detailed as possible nor as general as possible but that is, to my taste, just right (naturally).

The Quasi-static Approach

A quasi-static approach is one in which physical variables, although known to be subject to dynamic constraints, are treated as algebraic variables. In the physical sciences, this is a commonplace procedure. For example, the motions of the free ends of a lever are treated as if the motions of one end were literally simultaneous with the motions of the other end; inertia and transverse waves propagating along the lever are ignored. If a real lever is moved too rapidly, it will bounce off its fulcrum; one does not expect a quasi-static analysis to hold for such extreme cases.

The validity of the quasi-static approach as well as its usefulness depend on the frequency domain of interest. The designer of a man-machine system focuses on the high-frequency limits of performance because his task is not to understand the man but to get the most out of the machine for some extraneous purpose. This is the origin of the transfer function approach, and the reason why the engineering models can get away with treating the man in the system as an input-output box.

I am interested in the frequency domain that lies between a pure steady state and the "corner frequency," where the quasi-static analysis begins to break down. Thus, the analysis here does not encroach on the territory of engineering psychology. In the present analysis, there would be no point in carrying the transient terms of interest in engineering psychology because they go to zero before they become important. There would be a positive disadvantage in using mathematical forms that map the space being investigated into an intuitively unrecognizable space with non-physical variables in it ("cisoidal oscillations" or imaginary quantities found in Laplace transform theory and commonly applied to control systems; see Starkey, 1955, p. 31). The following analysis, while of little use for measuring transfer functions in the normal way, is suited to the elucidation of the structure of behavioral organization.

A Quasi-static Analysis

Consider a behaving system ("system" for short) in relationship to an environment. The system is the simplest possible: It has one sensory input affected by an input quantity, q_i, and one output that affects an output quantity, q_o. Both q_i and q_o are ordinary physical quantities in the environment or else are regular functions of measurable physical quantities.

In general, a change at the input to the system will result in a change at the output because of intervening system characteristics. The output quantity will be related to many other external quantities, but the only one of interest here is q_i, the input quantity. The input quantity will also be subject to disturbances from variables that change or remain constant independently of the output of the system.

The assumption of dynamic stability is made: After any transient disturbance, the system-environment relationship will come to a steady-state equilibrium quickly enough to permit ignoring transient terms in the differential equations that actually describe the relationship. This assumption implies the use of an averaging time or a minimum time resolution appropriate to each individual system.

It should not be thought that this assumption limits us to a static case. In the equation $F = MA$, or force equals mass times acceleration, the algebraic variable A is really the second derivative of position with respect to time. Nevertheless, there are many useful and accurate applications of this algebraic formula in dynamic situations. In a great variety of situations, time-dependent variables can be dealt with quasi-statically simply by a proper definition of the variables. All that is lost is the ability to predict behavior near the dynamic limits of performance in terms of the chosen variables. The system equation is

$$q_o = f(q_i), f \text{ being a general algebraic function.} \qquad (1)$$

(Small letters will be used for functions and capital letters for multipliers of parenthesized expressions when ambiguity is possible.)

The environment equation contains two terms representing linearly superposed contributions from two sources, which together determine completely the state of the input quantity. One contribution comes from the output of the system via q_o. The magnitude of q_o contributes an amount $g(q_o)$, where g is a general algebraic function describing the physical connection from q_o to q_i: This is the feedback path, which is missing when $g(q_o)$ is identically zero.

All other possible influences on the input quantity that are independent of q_o are summed up as an equivalent disturbing quantity, q_d, contributing to the state of q_i through an appropriately defined physical link symbolized as the function h; the magnitude of the contribution from disturbing quantities is thus $h(q_d)$. This provides the following environment equation (see Figure 3):

$$q_i = g(q_o) + h(q_d). \qquad (2)$$

The assumption of dynamic stability permits treating the system and environment equations as a simultaneous pair. To find a general simultaneous solution valid for all quasi-static cases in which physical continuity exists, we shall rearrange Equations 1 and 2 into equally general forms that are more manipulable. First, a Taylor series expansion of $f(q_i)$ is per-

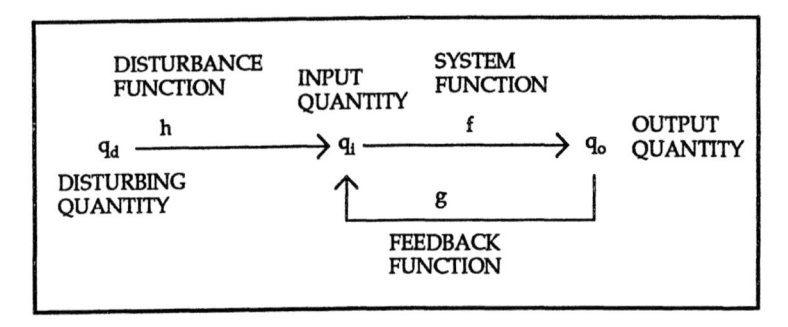

```
┌─────────────────────────────────────────────────────────┐
│   DISTURBANCE      INPUT         SYSTEM                 │
│   FUNCTION        QUANTITY       FUNCTION               │
│        h                            f         OUTPUT    │
│  q_d  ──────────▶  q_i  ──────────────▶  q_o  QUANTITY  │
│  DISTURBING                   ↑                         │
│  QUANTITY                     │     g                   │
│                               └─────────────┘           │
│                              FEEDBACK                   │
│                              FUNCTION                   │
└─────────────────────────────────────────────────────────┘
```

Figure 3. Relationships among variables and functions in the quasi-static analysis. (The topological similarity of Wiener's [1948] diagram, adapted in the present Figure 1, is of no significance because these variables and functions all pertain to observables outside the organism. This is not a model of the organism; it is a model of the organism's relationships to the external world.)

formed around a special (and as yet undefined) value, q_i^*, and an expansion of $g(q_o)$ is done about the corresponding special value q_o^*. For $f(q_i)$, the factor $(q_i - q_i^*)$ is factored out of the variable terms, leaving the following quotient polynomial:

$$q_o = f(q_i^*) + (q_i - q_i^*)[A + B(q_i - q_i^*) + C(q_i - q_i^*)^2 ...].$$

A, B, C, and so on are the Taylor coefficients. The quotient polynomial is symbolized as U to yield the following working system equation:

$$q_o = f(q_i^*) + U(q_i - q_i^*). \tag{3}$$

In a parallel manner, with the quotient polynomial symbolized as V, $g(q_o)$ is represented as $g(q_o^*) + V(q_o - q_o^*)$ to yield the working environment equation of

$$q_i = g(q_o^*) + V(q_o - q_o^*) + h(q_d). \tag{4}$$

Let the special value of q_i, or q_i^*, be defined as the value of q_i when there is no net disturbance: $h(q_d) = 0$. Then $q_i^* = g(q_o^*)$ and $q_o^* = f(q_i^*)$. Substitutions into Equations 3 and 4 then yield

$$q_o - q_o^* = U(q_i - q_i^*) \tag{5}$$

and

$$q_i - q_i^* = V(q_o - q_o^*) + h(q_d). \tag{6}$$

Substitution of Equation 6 into Equation 5 produces, after some manipulations twice involving the equivalence $V(q_o - q_o^*) = g(q_o) - g(q_o^*)$, Equation 7:

$$g(q_o) = q_i^* + \left(\frac{UV}{1-UV}\right)h(q_d), \text{ where } UV \neq 1. \tag{7}$$

Substituting from Equation 5 into Equation 6 directly yields Equation 8:

$$q_i = q_i^* + h(q_d)/(1-UV), \text{ where } UV \neq 1. \tag{8}$$

The dimensions of U are change of output per unit change of input, and the dimensions of V are change of input per unit change of output. Thus, the product UV is a dimensionless (and variable) number. It is customarily called the *loop gain* in morphologically similar equations of control theory.

So far these equations remain completely general, applying to any system-environment relationship of the basic form assumed, when the assumption of dynamic stability is observed to hold true. No model of the internal organization of the behaving system has been assumed, nor has it been assumed that we are dealing with a control system or even a feedback system. The only limits set on nonlinearity of the functions are practical ones: Systems that are radically nonlinear are not likely to meet the assumption of dynamic stability. These prove to be quite permissive limits.

Classifying System-Environment Relationships

The behavior of a system as defined here can be classified according to the observed magnitude and sign of the loop gain, UV. A severely nonlinear system can conceivably pass from one class to another during behavior.

Type Z: Zero Loop Gain

If the product UV is zero because the function f is zero, there is no behaving system. If it is zero because the function g is zero, there is no feedback and the simultaneous solution of the equations becomes (from Equations 1 and 2)

$$q_o = f(q_i) = f[h(q_d)].$$

This is the *open-loop* case and corresponds to the classical cause-effect model of behavior. If q_i is considered a proximal stimulus (located at the sensory interface or even at some stage of perceptual processing inside the system) and q_d a distal stimulus, then the output or behavior is mediated by the organism according to the form of the function f, and the proximal stimulus is the immediate cause of behavior. A stimulus object or event operates from its distal position as q_d, affecting the proximal stimulus q_i through intervening physical laws described by the function h. Thus, a simple lineal causal chain links the distal stimulus to the behavior.[1]

I shall say Z *system* to mean a behaving system in this Type Z relationship to its environment. In order to show that a given organism should be modeled as a Z system, it is necessary to establish that the organism's own behavior has no effect on the proximal stimuli in the supposed causal chain. I believe that this condition is, in any normal circumstance, impossible to meet. I will show later that even separating stimulus and response in time will not make the Z-system model acceptable.

Type P: Positive Loop Gain

If UV is positive and not zero, there is a Type P, or positive feedback, relationship between system and environment. The behaving system is then acting as a *P system*. This type of

---

[1]*Lineal* means occurring along a line or in simple sequence, as in lineal feet. *Linear* means described by a first-degree equation: For example, the equation $y = 3x$ expresses a linear relationship between x and y, while $y = 3x + 1/x$ expresses a nonlinear relationship.

Quantitative Analysis of Purposive Systems 143

relationship is dynamically stable only for $UV < 1$. A dynamic analysis is needed to show what happens for $UV \geq 1$; the algebraic equations give spurious answers. The P system goes unconditionally into self-sustained oscillations that either continue at a constant amplitude or increase exponentially or simply head for positive or negative infinite values of its variables. Whichever happens, the quasi-static analysis breaks down, as does the behavior of the system, since this is not generally considered normal behavior. A Type P relationship is dynamically stable only for $0 < UV < 1$.

There have been qualitative assertions in the literature that positive feedback may be beneficial because it "enhances" or "amplifies" responses. Such assertions are uninformed. Positive feedback does amplify the response to a disturbance because in a Type P relationship, behavior *aids* the effects of the disturbance on the input quantity. Equation 7 can be used to calculate the amplifying effects of various amounts of positive feedback, with the amplification factor being $UV/(1 - UV)$. The following list is an example of these calculations:

UV	Amplification Factor
.5	1.0
.6	1.5
.7	2.3
.8	4.0
.9	9.0
.99	99.0
≥ 1.0	unstable

In a nonlinear system, UV varies with the magnitude of disturbance; furthermore, natural systems have muscles that fatigue and interact with environments having variable properties. These facts are incompatible with the narrow range of values of UV (shown above), in which any useful amount of amplification is obtained from positive feedback. The relationship would always be on the brink of instability under the best of circumstances. We may expect natural P systems to be rare.

Type N: Negative Loop Gain

If UV is negative and not zero there is a Type N, or negative feedback, relationship between system and environment. The system is an N system. UV may have any negative value. In N systems, preservation of dynamic stability requires a trade-off between the magnitude of UV and the speed of response of the system. Servo-engineers would recognize the great advantage we have here over the person who has to design such a system: The designer has to tailor the dynamic characteristics to make the system-environment relationship stable; we only have to observe that it actually is stable. The equations we are using would be of no help to a designer of control systems.

It is difficult to find an example of behavior in which the feedback connection g is missing; feedback is clearly present in most circumstances. Moreover, it is generally found that organisms are sensitive to small changes of stimuli and that feedback effects are pronounced, so the magnitude of UV must be assumed in general to be large. Since we do not commonly observe dynamic instability, it follows that the sign of U must be opposite to that of V, that is, that the feedback is negative and the system an N system under most circumstances. Detailed investigation of individual cases, of course, will settle the question. I hope, however, that it can be seen that the Type N relationship is an important one. We shall examine its properties.

Properties of the Type N relationship. On the right side of Equation 7 is the expression $UV/(1 - UV)$, a form familiar in every mathematical approach to closed-loop analysis. With UV being dimensionless and negative for N systems, this expression is a dimensionless negative number between 0 and -1. Furthermore, the larger the minimum value of $-UV$ becomes, the more nearly $UV/(1 - UV)$ approaches the limiting value -1. When this limiting value is closely approached, we can call the system an *ideal N system*.

In the experiments to be described, the typical minimum value of UV estimated from the data was -30. Thus, only a 3% error is entailed in saying that subjects behaved as ideal N systems, in which $(-UV)$ is extremely large.

For an ideal N system, Equations 7 and 8 reduce to especially simple forms:

$$g(q_o) = q_i^* - h(q_d) \tag{7a}$$

and

$$q_i = q_i^*. \tag{8a}$$

From these equations can be drawn two basic statements that characterize a wide variety of N systems but more accurately for those that approach the ideal N system. Equation 7a is easier to translate if we remember that $q_i^* = g(q_o^*)$. The term $g(q_o)$ represents the effect of the output on the input quantity, $h(q_d)$ represents the effect of the disturbance on the input quantity, and q_o^* is the value of the output when there is no disturbance acting. It follows that the change in the output quantity away from the no-disturbance case is just what is required to produce effects on the input quantity that cancel the effects of the disturbance. Equation 8a expresses the consequence of this cancellation: The input quantity remains at its undisturbed value, q_i^*. Thus, the actions of an N system, mediated by the feedback path, stabilize its input quantity against the effects that disturbances otherwise would have. An ideal N system does this perfectly.

It will be seen that the widespread notion that negative feedback systems control their outputs is a misconception. In an artificial control system designed to produce outputs of interest to a user, the feedback function g is selected to make sure that q_i is precisely related to some objective consequence of q_o, so that controlling q_i will indeed result in controlling the objective consequence of q_o; however, such systems are built to protect themselves from all disturbances that might affect q_i directly. The erroneous transfer of an engineering model directly into a behavioral model was the cause of the misconception. The engineering model would show a reference input to the system, the effect of which would be to adjust the setting of q_i^* and also to indirectly affect the objective consequence. As mentioned, no such input from the outside exists in natural N systems (in none of those, at any rate, that I have investigated).

A behavioral illusion. Solving Equation 7a for q_o produces

$$q_o = g^{-1}[q_i{}^* - h(q_d)].$$

Compare this form with the equation for a Z system:

$$q_o = f[h(q_d)].$$

The difference in sign is a matter of choice of coordinates, and the constant $q_i{}^*$ can be used as the zero of the measurement scale, so the forms are essentially the same. The primary difference is that the organism function f in the Z-system equation is replaced by the inverse of the feedback function g^{-1} in the N-system equation.

This comparison reveals a behavioral illusion of such significance that one hesitates to believe it could exist. If one varies a distal stimulus q_d and observes that a measure of behavior q_o shows a strong regular dependence on q_d, there is certainly a temptation to assume that the form of the dependence reveals something about the organism. Yet, the comparison we have just seen indicates that the form of the dependence may reflect only properties of the local environment. The nightmare of any experimenter is to realize too late that his results were forced by his experimental design and do not actually pertain to behavior. This nightmare has a good chance of becoming a reality for a number of behavioral scientists. An example may be in order.

Consider a bird with eyes that are fixed in its head. If some interesting object, say, a bug, is moved across the line of sight, the bird's head will most likely turn to follow it. The Z-system or open-loop explanation would run about like this: The bug's position, the distal stimulus, is translated by optical effects into a proximal stimulus on the retina, exciting sensory nerves and causing the nervous system to operate the muscles that turn the head. This causal chain is so precisely calibrated and its form so linear that the movement of the head exactly compensates for the movement of the bug. The image thus stays centered on the retina.

There is a reason why this kind of explanation skips so rapidly across the proximal stimulus. If the head tracks the bug perfectly, the image of the bug will remain stationary on the retina, as indeed it very nearly does. But if the image remains stationary or wanders unsystematically about one point, the causal chain cannot be followed through. The open-loop explanation contradicts itself.

If the angle of the head is q_o and the visual angle of the bug is q_d, q_o has precisely as much effect as q_d has on q_i, the position of the retinal image. By choosing units properly, therefore, we can say that both g and h are unity multipliers of opposite sign. The two functions reflect the laws of geometric optics and, hence, are exquisitely precise and linear.

Equation 7a predicts that for an ideal N system, the output will vary as the inverse g function of the effect of the disturbance. Thus, the relationship between q_d and q_o will be as precise and linear as the laws of geometric optics. The organism function f, on the other hand, may be both nonlinear and variable over time. As long as the polynomial U remains large enough, the apparent behavioral law will be unaffected.

Thus, in the relationship between bug movement and head turning, we are not seeing the function f that describes the bird; instead, we are seeing the function g that describes the physics of the feedback effects. This property of N systems is well known to control engineers and to those who work with analog computers. It is time behavioral scientists became aware of it, whatever the consequences.

Operant Conditioning

The quasi-static analysis works quite well in at least one kind of operant-conditioning experiment: the fixed-ratio experiment, in which an animal provides food for itself on a schedule that delivers one pellet of food for each n presses of a lever.

The function g becomes just $1/n$, and there is no disturbance $[h(q_d) = 0]$.[2] The environment equation reduces to

$$q_i = q_o/n. \tag{9}$$

The average rate of reinforcement is treated as being the input quantity. The equations for an ideal control system predict

---

[2]Excessive efforts, on extreme schedules, would introduce a disturbance.

that $q_i = q_i^*$, which is to say that the organism will keep the average rate of reinforcement at a level q_i^* that is determined by a property of the organism. The average rate of bar pressing, using Equation 7a, will be $q_o = nq_i^*$.

From Equation 7a, we can predict what will happen if the schedule is changed from n_1 to n_2. The corresponding rates of bar pressing, q_{o1} and q_{o2}, will be related by

$$q_{o1}/q_{o2} = n_1/n_2. \tag{10}$$

The more presses are required to deliver one pellet, the more rapidly, in direct proportion, will the animal work the lever. This is a well-known empirical observation, found while "shaping" behavior to very high response rates.

A disturbance could be introduced by adding food pellets to the dish where pellets are delivered by the lever pressing at a rate q_d (the function h is then 1). The environment equation would then be

$$q_i = q_o/n + q_d. \tag{11}$$

From the solution for an ideal N system, we find

$$q_o = n(q_i^* - q_d). \tag{12}$$

If q_i^* is the observed rate of reinforcement in the absence of the disturbance, the rate of lever pressing in the presence of arbitrarily added food can be predicted. The average pressing rate will drop as the rate of adding food rises. When food is added arbitrarily at a rate just equal to q_i^*, lever pressing will just cease. This prediction is in accord with scientific observation (Teitelbaum, 1966) and with the qualitative empirical generalization that noncontingent reinforcement reduces behavior.[3]

It is evident that in order to predict quantitatively the results of this kind of operant-conditioning experiment, all one needs to assume about the organism is that it is an ideal N

---

[3]This is an excellent experimental method for measuring q_i^* in a natural situation.

system. The value of q_i^* can be found with one observation, and a whole family of relationships can be predicted thereafter. Conversely, the information obtained about the organism in such an experiment is only that it does act as an ideal N system controlling the rate of reinforcement. This analysis, while not dealing with learning, shows that changes of behavior do not necessarily imply any change of behavioral organization.

Let us now turn to a second quantitative method, which will be discussed more briefly but needs to be discussed because it deals with time delays, which the quasi-static approach cannot handle.

A Time-State Analysis with Dynamic Constraints

One persistent and incorrect approach to feedback phenomena is to treat an organism as a Z system, with any feedback effects being treated as if they occurred separately, after one response and before the next, thus apparently permitting the system itself to be dealt with in open-loop fashion. Qualitatively, this seems to work; but, as in every open-loop analysis, the approach fails quantitatively. The knowledge-of-results or stimulus-response-stimulus-response... analysis seems to succeed only because of the limitations of verbal or qualitative reasoning.

I shall use a linear model here, so I can focus on the main point without excessive complication. Let us alternate between the organism and the environment, first calculating the magnitude of the output quantity that results from the current magnitude of input and then calculating the next value of the input from the value of the output and the magnitude of the (constant) disturbance. This procedure leads to two modified equations. The system equation will be

$$q_{o(t+1)} = F(q_i - q_i^*)_t, \qquad (13)$$

and the environment equation will be

$$q_{i(t)} = Gq_{o(t)} + Hq_d. \qquad (14)$$

150 *Living Control Systems*

The functions *f*, *g*, and *h* have been translated into linear multipliers *F*, *G*, and *H*; and a time index, *t*, has been introduced. The loop gain is now the product *FG*, which corresponds to *UV* previously.

To skip a useless analysis, I will report that this set of equations converges to a steady state with *FG* in the range between +1 and -1 but not at or outside those limits. With the loop gain *FG* limited to *FG* > -1, the behaving system certainly cannot act like an ideal N system. The permissible amount of feedback is so small that there would be little behavioral effect from having any at all (except possible proneness to instability).

The difficulty here is that a sequential-state analysis of this kind introduces time without taking into account phenomena that depend on time. In the design of logic circuits, this can perhaps be done successfully, although a tight design has to recognize the fact that so-called "binary variables" in a logic network are really continuous physical quantities that, like any quantities in the macroscopic world, take time to change from one state to another. Ones and zeros exist only in abstract machines.

Without getting into a full dynamic analysis, we can introduce a dynamic constraint on this system by allowing the output to change only a fraction of the way from its current value of $q_{o(t)}$ toward the next computed value of $F(q_i - q_i^*)_{(t)}$ during the time between one value of *t* and the next. Letting *K*, a number between 0 and 1, be this fraction, let us introduce a modified system equation with this dynamic constraint:

$$q_{o(t+1)} = q_{o(t)} + K[F(q_{i(t)} - q_i^*) - q_{o(t)}]. \qquad (15)$$

Substituting Equation 14 into Equation 15 now yields

$$q_{o(t+1)} = q_{o(t)} + K[FGq_{o(t)} + FHq_d - Fq_i^* - q_{o(t)}],$$

or

$$q_{o(t+1)} = q_{o(t)}(1 + KFG - K) + KF(Hq_d - q_i^*). \qquad (16)$$

A steady state for repeated calculations with Equation 16 will be reached in one jump if (1 + *KFG* - *K*) = 0, implying an optimum value for K of

$$K_{opt} = 1/(1 - FG). \tag{17}$$

Setting $q_{o(t+2)} = q_{o(t)}$ leads to the value of K at which the system just goes into endless oscillation, that is, the critical value or upper limit of K:

$$K_{crit} = 2/(1 - FG) = 2K_{opt}. \tag{18}$$

If $K_{opt} < K < K_{crit}$, the successive iterations of Equation 16 oscillate above and below the steady-state solution, converging more and more rapidly as K approaches K_{opt}. For $0 < K < K_{opt}$, the successive iterations approach the same steady state but in an exponential monotonic way. In any case, the steady state (ss) is that found by substituting K_{opt} into Equation 16, with the result

$$q_{o(ss)} = \left(\frac{FG}{1 - FG}\right)\left(\frac{Hq_d}{G} - \frac{q_i^*}{G}\right). \tag{19}$$

For $FG \ll -1$ (an ideal N system), the expression $FG/(1 - FG)$ can be replaced by -1 to yield

$$Gq_{o(ss)} = q_i^* - Hq_d. \tag{20}$$

Comparing this with Equation 7a,

$$g(q_o) = q_i^* - h(q_d),$$

one can see that the linear sequential-state analysis with a dynamic constraint provides the same final picture of behavior that the quasi-static analysis provides. Treating behavior as a succession of instantaneous events propagating around a closed loop will not yield a correct analysis, no matter how tiny the steps are made, unless this dynamic constraint is properly introduced. With the dynamic constraint, the discrete analysis shows that behavior follows the same laws of negative feedback whether the feedback effects are instantaneous or delayed. This consideration has not, to my knowledge, been taken into account in other discrete analyses of behavior.

I have found this iterative approach useful in constructing computer simulations; even for highly nonlinear systems, it is usually possible to find a value of K that will stabilize the model. The behavior is essentially that of a system with a first-order lag.

Experimental Demonstration of Principle

Let us now look at six experiments that bring out fundamental aspects of this approach. They are not thought experiments, although I will describe them only in general terms; they were done with an on-line computer system using real subjects. The aim of the experiments was not to begin serious explorations of human nature using these organizing principles; that task lies in the future (and I hope I will not be the only one involved in it). The purpose of this effort has been to select from among dozens of experiments tried over the past 3 years a few that are easily replicated by many means that produce reliable results that can be explained only by the version of control theory used here and that always give accurate quantitative results, as good as those obtained in the laboratory demonstrations in introductory physics courses. Of course, the point of these experiments will be lost if nobody else tries them.

There is a way to tell when one has thoroughly understood each experiment and has discarded all inappropriate points of view. This is to persist until it is seen exactly why each quantitative result occurs as it does. When one realizes that no other outcome is possible for an ideal N system, one fully understands the experiment and also how an ideal N system works. To communicate that kind of understanding is what I hope for here.

General Experimental Method

A practiced subject sits facing a cathode-ray tube (CRT) display while holding the handle of a control stick that is pivoted near the subject's elbow. The angle of the control stick above the horizontal is considered the positive direction of behavior, below being the negative, and a digitized version of that measure in the computer is defined as the output quantity q_o.

On the screen is a short horizontal bar of light that can move up and down only over a grid of dots that remains stationary, providing a reference background. The position of the bar, or cursor, above or below center is taken as the input quantity, its measure being the digital number in the computer corresponding to displayed position q_i. This remotely defined q_i is valid because there are no disturbances intervening between subject and display that could alter the perceived figure-ground pattern.

Inside the computer is a random-number routine that repeats only after 37,000 hours of running time. Another routine smooths this random number, limiting its band width to about .2 Hz. The resulting number is the disturbing quantity q_d. The subject has no way to sense the magnitude of q_d directly.

The position of the cursor is completely determined at every instant (that is, 60 times per sec) by the sum of q_o and q_d. When all quantities are expressed in terms of equivalent units on the screen, the environment equation corresponding to Equation 1 is

$$q_i = q_o + q_d \qquad (21)$$

The system equation is just Equation 1: $q_o = f(q_i)$, where f is some general quasi-static algebraic function. The handle position is thus taken to depend in some way on the sensed position of the cursor.

A typical run begins with q_d forced to zero. During this time, the value of q_i is determined. By definition, this value is q_i^*. It is always measured, even though the instructions may appear to predetermine it; subjects do not always set q_i in the way the experimenter had in mind. A typical run lasts 60 sec after the random-number program is allowed to continue. The random-number generator runs continuously, but for the first part of each experiment, zero is substituted for the output of the smoothing routine. So far no two experimental runs have employed the same pattern of disturbances. If this seems like excessive zeal to attain randomness, it is done because a critic once suggested, apparently seriously, that the sine-wave disturbances I used at first were being memorized by the subject, even though there was no way for the subject to detect errors in phase or amplitude between the subject's actions and the changes in the disturbance and no way to sense the disturbance.

Experiment 1: Basic Relationships

The subject is requested to hold the cursor even with the center row of background dots (a standard compensatory tracking experiment). The value of q_i^* is determined as above, and the run commences. From Equations 7a and 8a, which presume that the subject is an ideal N system, it is predicted that $q_i = q_i^*$ and $q_o = q_i^* - q_d$. Here, q_i^* should be zero and very nearly is.

Figure 4 is a drawing of a typical result from a plot on the CRT screen. Any practiced subject will produce this kind of pattern.

The root mean square (RMS) variations of q_i about q_i^* (= 0 here) are about plus or minus 2% of full scale. Thus, q_o is, within the same tolerance, a mirror image of the disturbance q_d.

A minimum value of U can be estimated from a simulation that best fits the data. For most practiced subjects, it is at least -30 and may be much larger. The data suggest a first-order lag system (output proportional to integral of $q_i - q_i^*$), but no attempt was made to determine a valid transfer function. The assumption of stability is clearly met. There is little doubt that we are seeing a nearly ideal N system.

The best way to gain an intuitive understanding of Figure 4 is to start with the observed fact that the input quantity remains essentially at the value q_i^*. It follows that the handle must always be in the position that balances out the effect of the disturbance. We are not modeling the interior of the subject, so we need not worry about how this effect is created. It

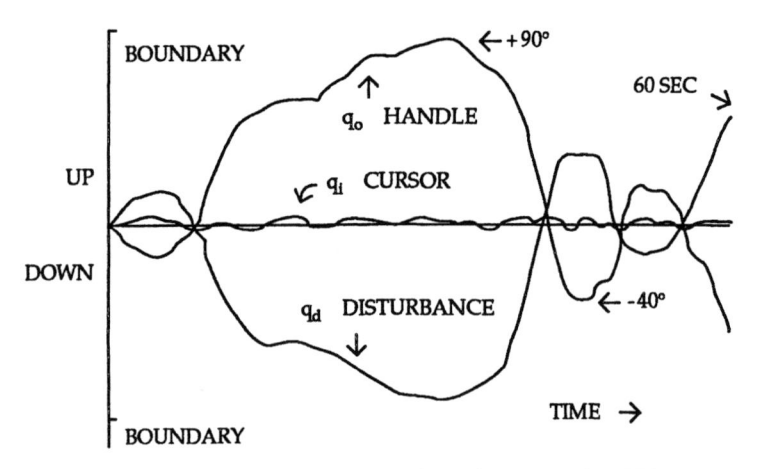

Figure 4. Experiment 1 results drawn from the cathode-ray tube (CRT) display of data. (The "cursor" trace represents the up-down position fluctuations that the subject sees on the CRT screen. The "disturbance" trace represents the invisible random quantity that is added to a representation of handle position ["handle" trace] to determine the position of the cursor.)

Quantitative Analysis of Purposive Systems 155

is a fact to be accepted. From the fact that the input is stabilized, the other relationships follow.

Experiment 2: Unspecified q_i^*

The subject is now asked to hold the cursor in "some other position," as accurately as possible. With $q_d = 0$, q_i^* is measured, and the run commences. The results are the same as before, with a nonzero value of q_i^*. This variation on Experiment 1 shows that the subject, not the apparatus or the experimenter, determines a specific quantitative setting of q_i^*.

Experiment 3: Change of Variable

The subject is asked to make the cursor move in any slow rhythmic pattern, the same pattern throughout the run. The subject indicates when the pattern on the screen (with $q_d = 0$) is the one to be maintained. The run commences. The initial pattern is taken to be q_i^*; there are many means for characterizing a temporal pattern quasi-statically, such as phase, amplitude, or frequency measures (one or more of which might prove to be controlled or uncontrolled). I used a much more subjective method, adequate for present purposes although not for serious work: eyeballing the data.

A typical result is shown in Figure 5. A separate plot is given for q_i to avoid confusing the curves. Without any disturbance, the measure of handle behavior is the same as the measure of the input quantity. Regularities in the cursor behavior appear to be just reflections of regularities in the handle behavior. When the disturbance is applied to the cursor, however, it is the handle behavior, not the cursor behavior, that begins to show corresponding large random fluctuations. This is not at all what the customary cause-effect model would predict.

Two major points are illustrated here. One is that more than one input quantity can be defined in a given experimental situation. The other is that the regularities we commonly term *behavior* are more likely in a natural environment to be associated with inputs than with outputs. Outputs reflect disturbances as well as the actions required to produce a given input

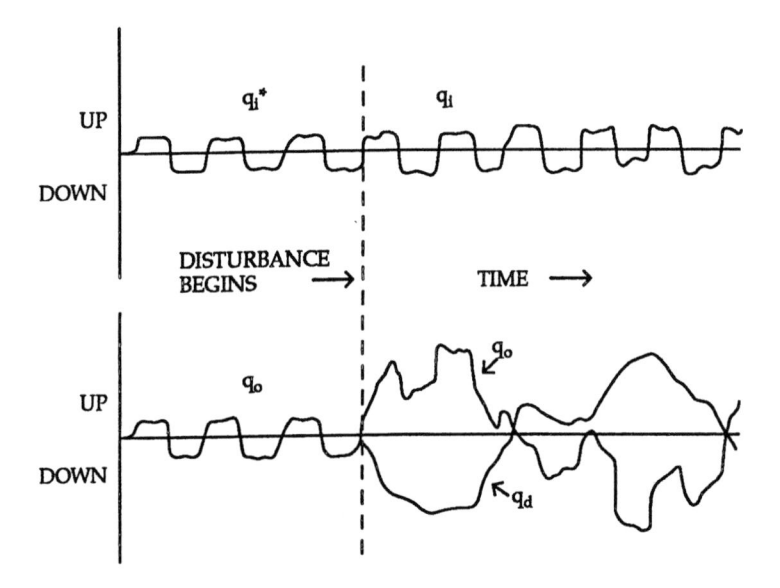

Figure 5. Experiment 3 results drawn from the cathode-ray tube (CRT) display of the data. (The upper trace shows the behavior of the cursor [q_i] on the CRT screen without [left] and with [right] the disturbance acting. The lower trace shows the handle position [q_o] with no disturbance acting [left] and the handle position [q_o] and disturbance magnitude [q_d] when disturbance begins changing [right]. The output, not the input, directly reflects the disturbance. The duration of the run was about 1 minute.)

pattern, and the component of output reflecting nothing more than disturbances may be by far the larger component. This fact may suggest why behavioral science so often has to rely on statistical methods to deal with its subject matter.

Experiment 4: The Behavioral Illusion

The conditions of Experiment 1 are now restored, and the computer is programmed to insert a nonlinear function between actual handle position and the effect of the handle on cursor position. This nonlinear function is the *g* function previously defined. Its form here is

$$g(x) = Ax + Bx^3.$$

The polynomial V is thus $A + Bx^2$. A and B are chosen so that the minimum value of V, at the center of the screen, is one third of the maximum value at the boundaries. If we call q_i^* zero, whatever its magnitude, and refer measures of q_i to that zero point, we can interpret Equation 7a to predict

$$q_o = g^{-1}[-h(q_d)].$$

Instead of computing the unwieldy inverse, we can simply plot q_o against q_d, for it is predicted that

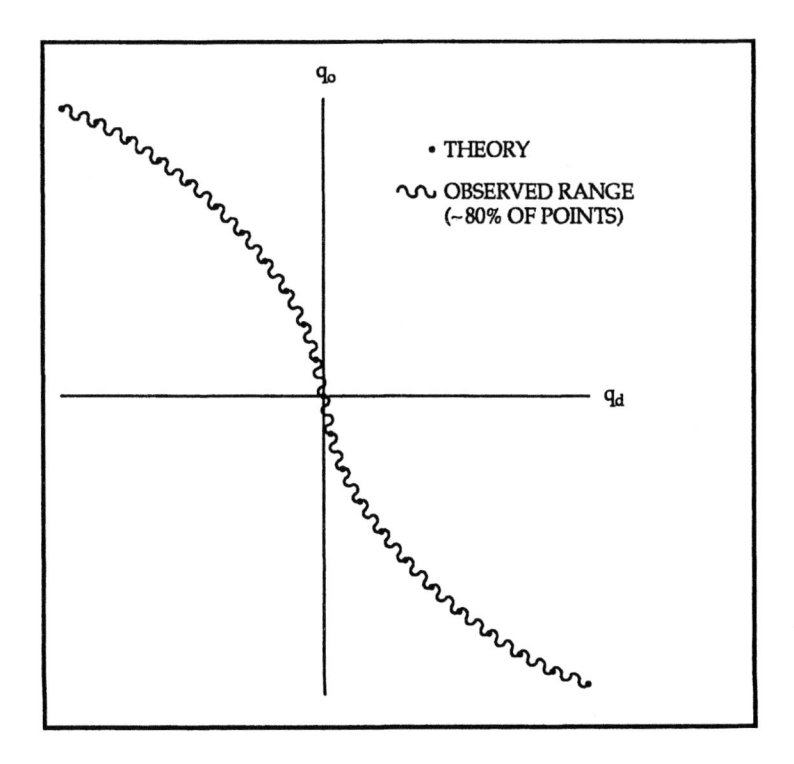

Figure 6. Experiment 4 results drawn from the cathode-ray tube display of the data, with a mildly nonlinear feedback connection. (Handle position is related to disturbance magnitude according to the inverse of the feedback connection. Dots represent the calculated inverse. Wavy lines show the approximate range of 300 data points for the practiced subject. The output quantity is represented by q_o. The disturbing quantity is represented by q_d.)

$$Aq_o + Bq_o^3 = -q_d, \text{ where } q_i^* = 0.$$

A typical result for any practiced subject is drawn in Figure 6. The RMS error between q_i and q_i^* remains about 2% of full scale. Most subjects notice nothing different about this rerun of Experiment 1. They are not paying attention to their outputs, except when actions become extreme because of a peak in the disturbance.

A more extreme version of the experiment involves choosing A and B to give the cubic form a reversal of slope near the center of the screen (see Figure 7). Most subjects do notice something different now: A few have complained that the handle is broken or that the computer is malfunctioning, although when they stop complaining they perform just as well as anyone else.

The reversal of slope converts the nominally Type N relationship to Type P near the center of the screen. Subjects simply skip across the Type P region as quickly as they can to the next stable point, where the feedback is once again negative. Over the rest of the range, the behavior is precisely what is predicted from the inverse of the g function.

A computer simulation using the successive-state method and a value of K chosen for stability behaves in just the same way, whether the behaving system is assumed linear or nonlinear. In fact, a three-level model I tried produced results indistinguishable from those for a real subject except for the very first move. The model had about 2% random noise in it.

The point of both versions of Experiment 4 is to show that the apparent form of the "behavioral law" connecting the distal disturbance to the behavior is determined strictly and quantitatively by the inverse of the feedback function and is, therefore, a property of the environment and not of the subject.

When these nonlinear feedback functions are used in Experiment 3, the subject still succeeds, although not as well, at maintaining a regular input pattern. A bystander entering at that point would have difficulty believing that the motions of the control handle had anything to do with the patterns on the screen. Yet the N-system equations sort out all effects neatly and quantitatively, with little random variation left over. It is all a matter of wearing the right pair of glasses.

Experiment 5: Multiple Choice

Now the display shows four side-by-side cursors instead of one, each moving up and down in its own band under the influence of its own independent random disturbance. The handle position contributes equally to the positions of all four cursors but affects Cursors 1 and 3 (C1 and C3) in the opposite direction to the effects on C2 and C4 (see Figure 8).

The subject is asked to pick any one cursor and hold it as steady as possible somewhere within its range of up-down

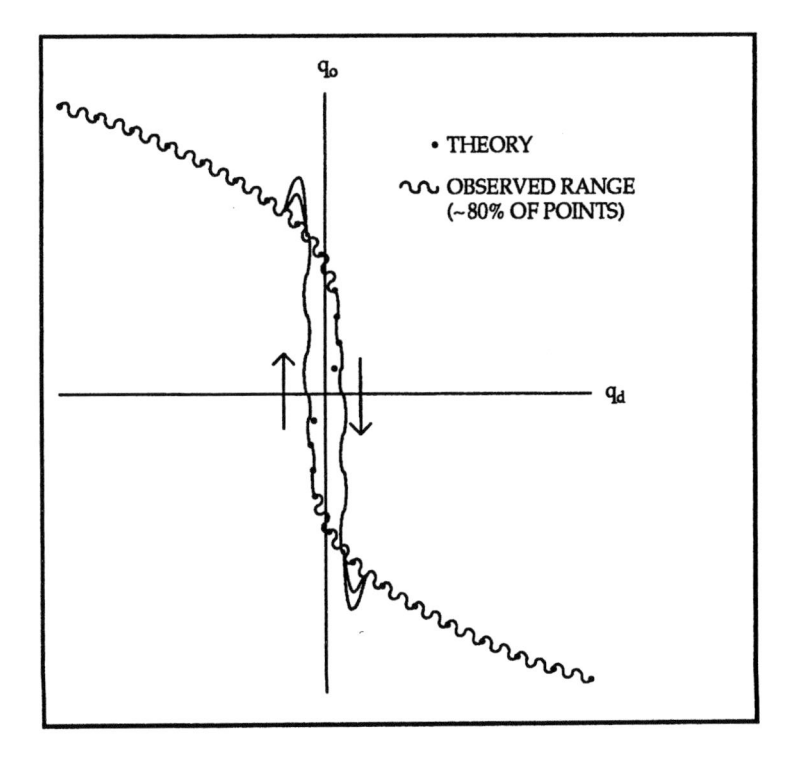

Figure 7. Experiment 4 results drawn from the cathode-ray tube display of the data, with an extremely nonlinear feedback connection (two-valued near center). (Subject's behavior [wavy lines] follows theoretical inverse, except near the center, where the region of positive feedback is skipped over. The output quantity is represented by q_o. The disturbing quantity is represented by q_d.)

160 Living Control Systems

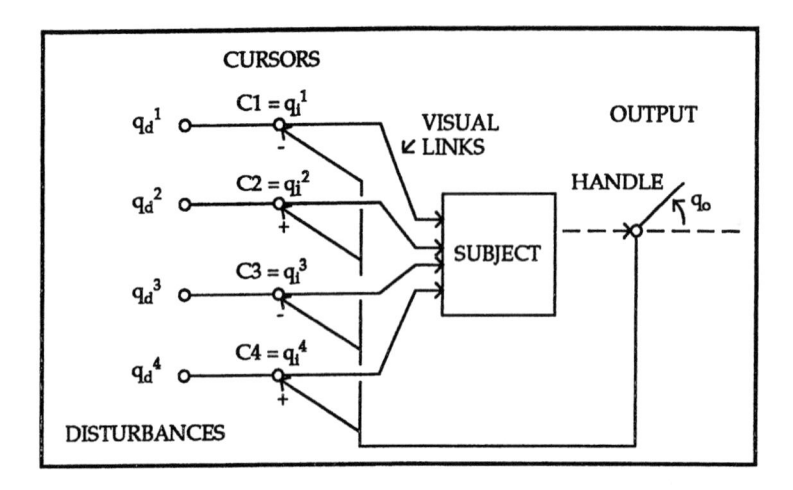

Figure 8. Analytical model for Experiment 5. (The subject sees four independently disturbed cursors [C1 through C4] on the cathode-ray tube. The handle affects all four cursors by an equal amount but in opposite directions for odd and even cursors. The subject can use the handle to control any one of [at least] 16 different aspects of the display.)

travel. The subject does so, with results indistinguishable from those of Experiment 2. One of the four cursors remains at the position q_i^* determined with all disturbances set to zero, while the other three cursors wander unsystematically up and down.

All cursors are input quantities; all are imaged on the subject's retinas. Only one, however, is a controlled input quantity. We can now distinguish controlled from uncontrolled input quantities and illustrate the *test for the controlled quantity*, which is a tool for investigating N systems of all kinds.

There are many possible variations of the test. One that works well for these experiments involves treating both handle movements and disturbances as random variables and comparing the expected variance, V_{exp}, of each controlled quantity with the observed variance, V_{obs}. Of course these variances must be calculated taking into account the hypothesized nature of the input quantity to be tested. The expected variance is computed by adding in quadrature the contributions from observed handle position and observed disturbances,

Quantitative Analysis of Purposive Systems 161

appropriately computed on the basis of analyzing the physical situation. Then, a stability factor S is calculated:

$$S = 1 - (V_{exp}/V_{obs})^{1/2}.$$

If $S = 0$, the input quantity is not controlled. If S is positive, behavior destabilizes the input quantity, and positive feedback exists. If S is negative, behavior stabilizes the input quantity, and negative feedback exists. For S several standard deviations more negative than -1, the behaving system can be called an ideal control system. For experiments like the first three, S is typically -4 to -9 for the controlled cursor, implying that the chances against an N system existing range from one in thousands to one in billions. For uncontrolled cursors, S ranges from +1 to -1 on short runs and comes close to 0 on long (10-minute) runs.

This statistical version of the test should be useful in cases where behavior takes place in a natural environment, where there are many possible effects of behavior, many sources of disturbance, and many potentially controlled quantities affected both by behavior and by disturbances. Once a controlled quantity has been found by this statistical approach, use can be made of the more quantitative methods of analysis previously discussed.

Any version of the test for the controlled quantity must be followed by verifying that an apparent controlled quantity must be sensed by the behaving system in order to be controlled. In the present experiments, covering up the appropriate cursor with a cardboard strip should, and does, cause the controlled quantity to become an uncontrolled one. Covering any or all of the other cursors has no effect at all.

Experiment 6: More-Abstract Controlled Quantities

Under the same conditions as Experiment 5, the subject is asked to hold constant some other aspect of the display (not specified by the experimenter) rather than the position of one of the cursors. Most subjects are initially baffled by this request, some permanently until given broad hints. Eventually, most see the possibilities of the fact that the handle affects odd and even cursors oppositely. One aspect is the difference in position between an odd and an even cursor. A subject can easily keep, say, C1 and C4 level with each other or C1 a fixed distance above or below C4. Both cursors wander up and down but always together. With suitable definitions, a controlled quantity can be found that unequivocally passes the test for the controlled quantity (four possibilities of this type exist).

Another type of controlled quantity is the configuration with three cursors lying along a straight line. Four possible controlled quantities of this type exist. Still another involves creating a fixed angle with one cursor centered at the vertex and the other two lying in the sides of the angle. All these are relatively easy to control once the subject has realized that they can be seen in the display. Only 1 of these 16 possible static controlled quantities can be controlled at a time because the control handle has only one degree of freedom.

What determines which controlled quantity will be controlled? The apparatus obviously does not, for it determines only the possibilities; not the behavior either—the output, with its single degree of freedom, affects all possible controlled quantities all of the time. The behaving system itself must be the determining factor. What the person attends to becomes the controlled aspect of the display. The person also determines the particular state of the selected aspect that is to serve as $q_i{}^*$. My efforts to make models of human organization have been aimed at explaining this type of phenomenon. It has been difficult at times to explain why such models are required when the listener is unaware that such phenomena exist.

In all these experiments, a typical correlation coefficient relating handle position to a noncontrolled quantity or its associated disturbance is in the range from 0 to .8. Statistics are poor in these short runs, but some correlations do occur even in long runs. The handle and the disturbances do affect the various cursors; correlations are to be expected there.

The correlation between a controlled quantity and either its associated disturbance or the handle position is normally lower than .1; a well-practiced subject will frequently produce a correlation of zero to two significant figures. At the same time, the correlation between magnitude of disturbance and handle position is normally higher than .99 (I can often reach .998 in the simpler experiments). To appreciate the meaning of these figures, one has to remember that the subject cannot sense any of the disturbances except through their effects on the input quantities, the cursor positions. If the controlled input quantity shows a correlation of essentially zero with the behavior, any standard experimental design would reject it as contributing nothing to the variance of behavior. But the disturbance that contributes essentially 100% of the variance of the behavior can act on the organism only via the variable that shows no significant correlation with behavior. Not only the old cause-effect model breaks down when one is dealing with an N system, the very basis of experimental psychology breaks down also.

Summary and Conclusions

I have examined in this article four mistakes that threw cybernetics off the track as far as psychology is concerned: (a) thinking of control theory as a machine analogy, (b) focusing on objective consequences of behavior of no importance to the behaving system itself, (c) misidentifying reference signals as sensory inputs, and (d) overlooking purposive properties of human behavior in man-machine experiments. Considering behavior, without going through any technological analogy, I have developed two mathematical tools for analyzing and classifying behaving organisms. The classical cause-effect model is included as a special case. Finally, I have introduced six experiments that illustrate classes of phenomena peculiar to control behavior and that cannot be explained under any paradigm but the control system model. (The last statement can be taken as a friendly challenge.)

I believe that the concepts and methods explored here are the basis for a scientific revolution in psychology and biology, the revolution promised by cybernetics 30 years ago but delayed by difficulties in breaking free of older points of view. Kuhn (1970) uses the term *paradigm* in the sense I mean when I say that control theory is a new paradigm for understanding life processes—not only individual behavior but the behavior of biochemical and social systems. Chapter X in Kuhn's book discusses "Revolutions as Changes in World View." The experiments we have seen here, while not of great importance in themselves, represent my attempt to show how control theory allows us to see the same facts of behavior that have always been seen but through new eyes, new organizing principles, and new views of the world of behavior.

The natural tendency of any human being is to deal with the unfamiliar by first trying to see it as the nearest familiar thing. That is what happened to the basic concepts of cybernetics. It will happen even more pronouncedly in response to the ideas we have looked at here. The difficulties faced by a new paradigm, as Kuhn explained so clearly, result not from battles over how to explain particular conceptual puzzles, but from bypassing altogether old puzzles that some people insist for a long time still need solving. There are still many fruitful areas of research and many unsolved problems concerning the properties of phlogiston. Modern observational and data-processing techniques in astronomy could lead to great (but unwanted) improvements in the predictive accuracy of the epicycle model of planetary motions (I knew a graduate student in astronomy who showed how well epicycles could work with the aid of a large computer).

Control theory bypasses the entire set of empirical problems in psychology concerning how people tend to behave under various external circumstances. One kind of behavior can appear under many different circumstances; instead of comparing all the various kinds of causes with each other while looking for objective similarities to explain the common effects, we are led by control theory to look for the *inputs* that are disturbed not only by the discovered causes but by all possible causes. For a thousand unconnected empirical generalizations based on superficial similarities among stimuli, I here substitute one general underlying principle: *control of input*.

References

Chapanis, A. Men, machines, and models. *American Psychologist*, 1961, 16, 113-131.

Kelley, C.R. *Manual and automatic control.* New York: Wiley, 1968.
Kuhn, T. *The structure of scientific revolutions.* Chicago: University of Chicago Press, 1970.
Maslow, A.H. *The farther reaches of human nature.* New York: Viking, 1971.
Mayr, O. *The origins of feedback control.* Cambridge, Mass.: MIT Press, 1970.
McDougall, W. The hormic psychology. In A. Adler (Ed.), *Psychologies of 1930.* Worcester, Mass.: Clark University Press, 1931.
Poulton, E. *Tracking skill and manual control.* New York: Academic Press, 1974.
Powers, W.T. *Behavior: The control of perception.* Chicago: Aldine, 1973.
Powers, W., Clark, R., & McFarland, R. A general feedback theory of human behavior. Part 1. *Perceptual and Motor Skills Monograph,* 1960, 11. (a)
Powers, W., Clark, R., & McFarland, R. A general feedback theory of human behavior. Part 2. *Perceputal and Motor Skills Monograph,* 1960, 11 (b)
Rosenblueth, A., Wiener, N., & Bigelow, J. Behavior, purpose, and teleology. In W. Buckley (Ed.), *Modern systems research for the behavioral scientist.* Chicago: Aldine, 1968.
Skinner, B.F. *The behavior of organisms.* New York: Appleton-Century-Crofts, 1938.
Skinner, B.F. Freedom and the control of man. In B.F. Skinner (Ed.), *Cumulative record.* New York: Appleton-Century-Crofts, 1972.
Starkey, B.J. *Laplace transformations for electrical engineers.* New York: Philosophical Library, 1955.
Teitelbaum, P. The use of operant methods in the assessment and control of motivational states. In W.K. Honig (Ed.), *Operant behavior.* New York: Appleton-Century-Crofts, 1966.
Tolman, E.C. *Purposive behavior in animals and man.* New York: Century, 1932.
Von Foerster, H., White, J., Peterson, L., & Russell, J. (Eds.). *Purposive systems.* London: Macmillan, 1968.
Wiener, N. *Cybernetics: Control and communication in the animal and the machine.* New York: Wiley, 1948.

[1979]
A Cybernetic Model for Research in Human Development

Over the past thirty years, cybernetics has gone in many directions. It is sometimes difficult to see how some modern approaches that go under that name relate to the original concept proposed by Norbert Wiener (1948): the concept that organisms display the characteristics of negative feedback control systems. Since this concept represented the chief revolutionary departure of cybernetics from conventional thinking, one might expect every person claiming to be a cyberneticist to understand the principles of control theory, at least at the level of valid rules of thumb. This is not the case. What most "cyberneticians" do and write is perfectly compatible with traditional models of organisms, and hence is incompatible with the principles of control theory. In this volume, we hope to improve on that state of affairs.

The original promise of a cybernetic revolution in our understanding of human nature can still, I am confident, be realized. To bring it about, however, we must be prepared to change some concepts that have been defended for a long time. We must also be prepared to return for a while to a relatively low level of abstraction, so as to grasp the meaning of control theory in relationship to simple direct experiences. The first step in launching a cybernetic revolution in psychology is to make sure that the fundamental phenomena of control are correctly understood in relationship to behavior.

In this chapter I will be presenting a primer in control theo-

Copyright 1979 by Westview Press. Reprinted with permission of the publisher from Mark N. Ozer, ed., *A Cybernetic Approach to the Assessment of Children*, 1979, 11-66.

ry for behavioral scientists. While this book is concerned with human development, I think most readers would agree that research concerning *how* human beings develop into adults must be strongly conditioned by *what* one considers to be the nature of an adult organization. A thorough grasp of the principles of control systems will provide a picture of development considerably different from what one would obtain from, say, Freudian theory or behaviorism. Near the end, I will take a few steps toward applying this theory specifically to development, but only as any theoretician points out general directions to experimentalists or clinicians.

Background

Controlling various aspects of experience is an activity that has engaged human beings throughout recorded history. Until the middle 1930s, however, no human being was able to say, in a systematic and quantitative way, just what the term "control" meant. Like digestion, control was a natural process of which we could take advantage without any understanding of how it works.

Otto Mayr's book, *The Origins of Feedback Control* (1970), is a history of all known control devices from the water-level regulator of Ktsebios in the third century B.C. to the automatic watch regulator of Bregeut near the beginning of the nineteenth century. It is a slim volume. It shows that control systems were, during this 2,100-year period, freaks in a world of "normal" machines designed according to completely different principles. Control devices were built, but control principles were unknown; there was no development of a body of knowledge that could be passed along the generations. The concept of mechanism that developed at the time of the Renaissance and led to the first Industrial Revolution was based on a particular concept of cause and effect, in which one set of circumstances led through rigid linkages to the next, in strict temporal and spatial sequence. This concept of mechanism lay at the foundations of all the new sciences: physics, chemistry, and biology. No scientist, no engineer, realized that in the control system a different arrangement of cause and effect existed.

In 1868, James Clerk Maxwell (1965) published the differential equations for Watt's flyball governor. By using differential equations, Maxwell employed a method that transcends simple cause and effect for two reasons. First, the variables used in differential equations involve time-derivatives, and in the solutions, time integrals, representations involving processes that span time. The solutions of a set of differential equations arise from causes occurring not at an instant, but over an interval. Second, the equations Maxwell used dealt with processes in a *closed loop*. By breaking the governor down into part-systems, representing each part by an equation, and then solving the equations as a simultaneous system, Maxwell effectively removed temporal sequence from the picture and substituted simultaneous interactions in a circle with no beginning and no end.

This development occurred just as experimental psychology was becoming a science in Germany. It led to a long chain of developments—but in mechanics, not in psychology. The implicit new concept of mechanism remained implicit, being unrecognized even by engineers. Psychology continued to employ the old concept of mechanism, so that when the American psychologist James Watson announced the birth of behaviorism just after the turn of the twentieth century, he rested his case, as biology had done before him, on the assertion that such mechanisms as Maxwell had analyzed could exist only in a metaphysician's imagination.

As far as Watson was concerned, and as far as any life scientists of his time were concerned, organisms could obey the laws of nature only in one way. They must be basically passive devices set into motion by external forces. However intricate their inner construction, they could do nothing of themselves. Aquinas, trying to prove the existence of God, had declared that nothing moves of itself. Those in the life sciences who were trying to prove the opposite said the same thing.

In reaction against the idea that the universe worked according to the whims of Divine Purpose instead of discoverable Natural Law, life scientists have always made the mistake of assuming that purpose is a metaphysical fancy. This prejudice prevented them from looking into the many instances of behavior in which purpose appeared obvious; rather than asking what kind of organization could select its own goals and act so as to achieve them, the great majority of life scientists agreed that this appearance must be an illusion. They set themselves the task, at least tacitly, of dispelling the illusion by constructing cause-effect explanations in which physics and chemistry alone lay at the base, and external events worked through known physical processes to produce the apparently purposeful behavior. This, of course, left subjective phenomena in the role of accidental side-effects, and removed from organisms all capability for directing their own actions.

This logical scheme entered psychology as the twentieth century began, but its roots were in biology, medicine, physics, and chemistry. There is, therefore, a tremendous weight of tradition behind it. Whole disciplines are founded on it. Knowledge built on knowledge is also built on the assumption, which we can now prove false, that organisms cannot have internal purposes. Removal of that assumption will have profound and, from one point of view, disastrous effects.

Control theory finally became a formal system during World War II. By that time engineers had succeeded in building devices that imitated the human ability to act on the outside world and to control numerous aspects of it, and had found the mathematics for analyzing and designing such systems. Many of them were aware of the similarities of control system behavior to that of living systems. Norbert Wiener and his neurologist friend Rosenbleuth may have thought they were discoverers of this parallel, but I suspect that their engineer-collaborator, Bigelow, had been aware of it for a longer time. Engineers, after all, were the ones who gave such names as sensor and comparator to parts of control systems and whose job it was to analyze human tasks for the purpose of designing machines that could take them over.

I think there is also evidence that neither engineers nor the new cyberneticists realized that control theory entailed a concept of mechanism that went beyond the concepts at the foundations of the life sciences. Most engineers are dyed-in-the-wool mechanists, trained in the same ideas of cause and effect that have always been accepted. In a famous debate (Buckley, 1968), Wiener did not make the crucial distinctions between a passive, nonpurposive system (a compass) and an active, purposive system (a homing missile). They could not throw off the old point of view sufficiently to see that they were themselves talking about a new one.

Thirty years have gone by since Wiener joined a scientific revolution and coined a name for it. The confusion may finally be clearing up, the false trails may be in the process of abandonment. What we have left is considerably simpler than what cybernetics has gone through in those decades, but it is also fundamental. The message that Wiener gave us is not as complex as he made it out to be, and it is more important than he realized when he gave it. Organisms are purposive. Purpose is not a metaphysical concept. Behavior is a link in a process by which organisms control the most important effects that the environment has on them.

Let us turn now to Psychology 101 (ca. 1988). We begin by looking at ordinary behavior to see how control phenomena appear to an observer. Then we look at an elementary model of organization that can account, at least in terms of one real possibility, for what is observed. After that, I will present a picture of possible levels of organization in a human hierarchy of control organizations, and at the end, at last, some remarks of how this whole approach might be applied to research in human development. Anyone writing at the early stages of a scientific revolution has to rely on reason and supposition more than on hard data, but I can trust that the readers of this volume will be able to distinguish well-founded principles from conjectures about their application.

Characteristics of a Control System

The Phenomenon of Control

Control phenomena can be found in any example of behavior, even the briefest snapshot of ordinary activities. Consider an episode that lasts perhaps thirty seconds. A man gets in his car, starts the engine, puts the car in gear, and backs out of the driveway. Table 1 is a partial listing of activities that are likely to occur during this brief event. Each activity involves four items, the first of which is in the Behavior column. In this col-

Table 1. Control actions involved in getting ready to back a stick-shift car out of a driveway.

Behavior	Means	Variable	Reference state
Open door	Grasp, pull	Angle of door	80 degrees
Get in	Bend, sit, slide	Relationship to seat	Seated
Shut door	Grasp, pull	Angle of door	0 degrees
Fasten belt	Push together	Distance between fasteners	Zero distance
Adjust rear view mirror	Grasp, twist	Displacement, rear window image	Zero displacement
Depress clutch	Push with left leg	Extension of leg	Fully extended
Insert key	Extend arm	Distance, key to keyhole	Zero distance
Start engine	Twist	Sound of starter and engine	Whirrrrrr, vroom!
Shift to reverse	Grasp, push	Position of shift lever	Coordinates of reverse gear

umn are phrases of a kind used both in ordinary discourse and in scientific psychology to denote what an organism is doing. Even though Watson (1919) wrote that behaviorists, when necessary, reduce any behavior described at this level to a collection of individual reflex movements, he never actually did so. Nor has anyone else since. This is the level at which most scientists perceive the actions of organisms. No analysis takes place: instead, instances of behaviors described this way are counted, and the sums are used in statistical manipulations. We will at least try to analyze, but not with the results anticipated by Watson.

Consider the first item: Open Door. The meaning is obvious. The door which was shut becomes open, and the driver did it. The problem here, as in so many areas of psychology, lies in unspoken assumptions and relationships taken for granted. The opening of the door is certainly an event, but this event can be produced in innumerable ways. The door is not part of the driver. It is an object that can be in a number of states, and its state is determined by the sum of all forces that act on it. To say that the door opens says nothing about the driver.

The opening of the door tells us only that sufficient force existed to make the door swing open. There are many forces normally acting on a car door. Any tilt of the car is translated, owing to gravity, into a force tending to open or close the door. If a wind is blowing, it too exerts forces on the door. The door is held fully open by a mechanical detent, which resists the effects of other forces up to a point. While the door is moving, friction in its hinges and inertial effects of its mass create more forces. All these effects are normally present no matter how the door is being opened.

To say that the driver opens the door means only that the driver adds forces of his own to others that are acting on the door at the same time. He does not open the door by means of opening the door. He does it by *grasping* the handle, and by *pulling* on it. The second column of Table 1 lists some of the means used by the driver in bringing about each "behavior." We can see immediately that the real actions of the driver—the effects which we can attribute to the driver alone—are all in the Means column. What is so casually called behavior results from the conjunction of many forces, only one of which is contributed by the driver. The Behavior column really lists *consequences* of the driver's actions, consequences that are not determined by the driver's actions, but are only influenced by them.

By paying careful, even compulsive, attention to detail, we have uncovered a problem. What is casually called behavior, the opening of the door, is not something that could be attributed to the driver's actions alone. It results from the conjunction of many effects, only one of which is contributed by the driver. This problem exists for every entry in the Behavior column, and in fact for nearly any kind of behavior that can be mentioned, for any organism from the bacterium to the human being. "Behavior" is really a consequence that results from adding together many influences, the majority of which act even in the absence of the behaving organism. To speak this loosely of behavior, therefore, is to gloss over a fundamental problem that must be solved before any theory of behavior can possibly make sense.

Instead of standing back and noting that in a fuzzy sort of way the driver's action "opens the door," let us continue to pay attention to detail, asking *how* this behavior unfolds. The phrase "opens the door" refers to an amalgam of ongoing processes, mixed with implication about starting and ending conditions; it is sloppy, as most ordinary language is. In ordinary discourse, we can often straighten out the messge as we receive it, but in science that sort of informality is worse than useless.

The driver's action, combined with other forces, does not simply conjure up an event out of nothing. It produces a process that goes smoothly from a starting condition to a final condition. At every moment during this process, there is a *variable*, the angle made by the door and the car frame, that has some particular value and some rate of change. This variable, and not "the door," is what changes, what is influenced by all the forces being applied. (Column 3 of Table 1 is a listing of some of the variables associated with each behavior.) We can represent the bulk of what is meant by opening the door in terms of the way this variable changes from one value (0 degrees) to another (say, 80 degrees), going through all the intermediate angles.

The driver will pull on the door hard enough to overcome the resistance of the door, counteract wind and gravity forces, and start the door swinging open. He will stop pulling when the door is open, meaning that the angle has become 80 degrees. We can confidently expect that this will occur no matter what the wind or the tilt of the car, and no matter what car is involved. It does not matter that on one occasion the driver may have to pull with a force of 1 pound and on the next (with a different car, wind, and tilt) a force of 50 pounds. The driver will simply pull *hard enough* to produce the result we expect.

So, by being patient we have uncovered a second major

problem. What is uniform about this behavior (or any of the behaviors listed) is that it occurs regardless of even large variations in extraneous factors that contribute to the final result. The action in the Means column changes from one occasion to the next in just the way needed to make up for changes in other contributions. The variations in action are not small; if the wind blows hard enough to propel the door by itself, the driver might open the door by *pushing* (after it is partly open).

Column 4 of Table 1, Reference State, refers to the final condition to which the variable is brought despite any ordinary disturbance. The existence of these reference states is not conjectural; once behavior has been defined in terms of an appropriate variable, such reference states always exist. They can be discovered experimentally, and defined in terms of observable relationships. Whether or not they *should* exist according to anyone's theory, they *do* exist.

In these reference states we have the heart of the problem to which control theory is addressed. What kind of system can behave in such a way that a variable will, under a variety of unpredictable conditions, always approach the same state? What determines that state? *Where* is that state determined —that is, by what? To dismiss the existence of reference states as an illusion is simply to discard data. To explain such reference states as the inevitable outcome of prior causes in the outside world is to demand experimental verification of a kind that has never been found. To say that "subtle cues," which are so subtle as to escape the notice of a careful observer, cause the singularly appropriate variations of action is to abandon science.

Reference states cannot exist under the old cause-effect model. They refer, as far as external observations are concerned, only to future states of the organism or its environment. They cannot affect present behavior, and they must be treated simply as *outcomes* of events caused by prior events. The flaw in this reasoning is hard to see if one does not know (as the founders of scientific psychology did not know) of organizations capable of complex internal activities that are essentially independent of current external events. By ruling out the possibility of significant causes of behavior *inside* the organism, where they could not be observed, early behavioral scientists in effect committed themselves to a whole chain of deductions following from the assumption that everything of significance with regard to behavior could be observed from outside the organism. They were betting everything on the assumption that such internal causes would never be found to exist, partly because the methods they used could not be valid if such inner causes *did* exist. So when we speak of reference states here, we are resurrecting a corpse that was buried a long time ago and which is still preferred by many to be left underground. Only control theory justifies this disinterment.

There is one explanation for the existence of reference states that has been proposed over and over for centuries: they are determined by the *intentions* of the behaving organism. The driver has, inside him, the intention that the door be open. He acts to achieve this purpose, doing whatever is required (if possible) to achieve it. This simple and parsimonious explanation has only one fault, or had prior to control theory: intentions are "mental" phenomena, and here we are asking them to have "physical" effects. This apparent difficulty has stopped scientific psychology in its tracks for some 75 years if not more.

There is another approach to the problem. Instead of automatically assuming that mental and physical phenomena have nothing to do with each other, we can assume that there is no contradiction and try to find out how this result is brought about—how the phenomenon of inner purpose or intention works. We do not have to accept the idea that the future can affect the present, nor do we have to repudiate any useful principles of physics and chemistry that have been so carefully constructed. All we have to do is find an organization that can do what we observe being done. That is what we shall now do.

The Organization of a Control System

For *any* of the "variables" of Table 1 to be brought to specific reference states, the driver must be able to sense them —either directly, or by sensing something that covaries with the variable we choose to measure the behavior. Elementary experiments would be enough to establish this principle. If the driver had to execute any of the behaviors in Table 1 blindly, with no visual, auditory, kinesthetic, or other sensory information to tell him the current status of the variable, it would be impossible for him to vary his actions so as to oppose unexpected disturbances. In fact, we would find through continuing experiment that the only reliable consequences of the driver's actions are those the driver can sense. This is a crucial hint about how this sort of phenomenon is created.

It is necessary for the driver to sense the controlled variable, but not sufficient. Suppose the driver senses the position of the gearshift lever. He senses the position known as first gear. Does this imply *pull* or does it imply *push*? By itself, the information given by the position of the shift lever does not imply *any* action. It merely indicates the current state of affairs. The driver needs to know where the lever *is to be*, not just where it is.

By the same token, it is not sufficient for the driver to know where the lever *is to be*. If the lever is to be in the neutral position, the action required to put it there depends on where the lever *is*. No amount of description of the reference condition can, by itself, lead to an action that will make the controlled variable approach it.

Creating an action which will bring the controlled variable to some specific reference condition depends on information contained partly in the present state of the variable and partly in some specification of its reference condition. The direction and amount of action that is required depends on the direction and amount of discrepancy between these two quantities. Therefore, in order for this discrepancy to be corrected reliably, the driver must be organized to detect its amount and direction and to convert it into those coordinated acts that will systematically reduce the discrepancy. A decreasing discrepancy should lead to a decreasing amount of action, and a reversal of the discrepancy should lead to a reversal of the action. A change in direction of the discrepancy should produce a corresponding change in direction of the action, so the action is always aimed against the direction of the discrepancy.

We have, therefore, what amounts to a design problem. It is instructive to see how the engineers of the 1930s solved it, being unaware that purposes or goals were metaphysical constructs.

Suppose an angular position like that of the car door is to be controlled. The first step is to provide a means of *sensing* it. To the engineer, this means converting the angle of the door into something with which an electronic circuit can deal, say a voltage. This can be done easily by attaching a transducer to the door hinge, a transducer that generates an output of 80 volts when the angle is 80 degrees and 0 volts when the angle is 0 degrees, and that varies between these voltages as the door varies its position between those angular limits. This sort of representation is known as *analog* representation. The signal generated by the transducer does not change position or angle as the door does, but each *magnitude* of the signal corresponds to one *position* of the door.

The same system would work in a brain. As the image of the car door on the driver's retina changes, and as other kinds of information change, some signal in the driver's brain varies in magnitude in a corresponding way. Magnitudes in a neurological model correspond to frequencies of firing. Of course more than simple sensing is required; the uncountable signals from many kinds of sensory receptors must be combined through real-time neural computations to generate a position signal like that from the transducer.

The transducer signal merely indicates where the door is, not where it is to be. Since the position of the door now exists as a voltage analog, a particular position corresponds to a particular voltage. It is easy to arrange for a second source of volttage, independent of the transducer, which can be set to create any voltage between 0 and 80 volts. This voltage would be called the *reference signal*. If the reference signal were set to 65 volts, then if the transducer signal were somehow made to match it, we could deduce that the door must stand at an angle of 65 degrees. Thus setting the reference signal to some specific voltage is equivalent to specifying some particular state of the variable that is being sensed—the angle of the car door.

If we can imagine a neural signal that is the brain's analog of the car door angle, it is certainly easy to imagine another neural signal, independent of the sensory signals affected by the position of the door, which is fixed in magnitude between the magnitude limits of the sense-derived signal. If, somehow, the sensory signal could be made to match the neural reference signal in magnitude, the angle of the door would be indirectly specified by the setting of the reference signal.

In the electronic control system it is a simple matter to subtract one signal from another, leaving a difference signal. This is done with an electronic *subtractor*, or, as it is known in control applications, a *comparator*. This is a device which responds only to the algebraic (signed) difference between two voltages and generates an output voltage whose magnitude depends on the magnitude of the difference, the sign depending on the sign of the difference (voltages can be either positive or negative). This output of a comparator is called by engineers an *error signal*.

Subtraction can also be accomplished with neurones. If two neural signals combine at one nerve cell, one may act to excite firings of that cell while the other acts to inhibit firings of that cell. Both excitation and inhibition take place in a smooth, graded manner, provided one treats the phenomena in terms of frequencies of firing and neurotransmitter concentrations, not single impulses. The frequency with which the receiving cell fires then depends on the *difference* in frequencies between an exciting and an inhibiting input. Frequencies cannot go negative, but if a second cell existed in which the roles of excitation and inhibition by the two incoming signals were interchanged, the two cells together could generate a pair of error signals, one representing "too large" and the other representing "too small." Absence of both signals would mean that the sensory input signal exactly matches the reference input signal.

To complete the artificial car door control system, the engineer needs only to convert the electronic error voltage into an effect on the car door. A positive voltage should act to increase the angle, a negative voltage to decrease it. This is accomplished by letting the error voltage enter a power amplifier, which boosts the level of energy from that of an information-carrying signal to that of a power source capable of running a motor, still retaining the basic directionality and magnitude information in the error signal. If a positive voltage means that the transducer signal is less than the reference signal and turns the motor to *open* the door, and if a negative voltage means the opposite and causes the opposite direction of turning of the motor, the door will be urged toward a reference state whether it is initially too far open or too far shut.

It is the same in the brain. If the pair of error signals is converted either into a push or a pull depending on the sign of the error (i.e., which of the pair of error signals is present), and if the magnitude of the effort depends on the magnitude of the error, the car door will always be urged toward the position in which it would generate a sensory signal equal to the reference signal.

The engineers who devised systems that could imitate human control behavior proved, by the expedient of building examples of their model, that this model of human organization does in fact work. Of course many other organizations would also work, since there are many ways to sense an angle, and many more complex ways of converting an error into an action that will correct an angle error. But all such alternate systems are equivalent to the organization we have just been through, and none of them is equivalent to the simple cause-effect model that tradition assumes. Thus the existence of working artificial control systems supplies us with an existence theorem, proving that a real physical system can act to make a variable approach a preselected state with no metaphysical assumptions at all being needed. Given that proof, we do not have to worry about whether the model we adopt is exactly the correct one or not. We know we have the right *kind* of model and can commence finding the correct one of that kind through the usual scientific procedures of hypothesis-testing.

Perhaps now the difference between the old cause-effect linear model and the control-system model is becoming more apparent. Under the old model, stimuli were thought to act on the nervous system to cause muscle tensions; those muscle tensions caused the limbs to move, and those motions in turn created all the effects we call behavior. It was assumed that the final patterns were simply the sums of all the detailed muscle tensions, added according to the laws of vector addition and creating consequences according to the laws of mechanics. If disturbances entered in such a way as to cause changes between the time of the generation of muscle tensions and the later consequences of those tensions, one could not predict logically that the consequences would remain the same.

Control theory predicts that the consequences—or at least certain of them—*will* remain the same despite such disturbances. They will remain the same because the organism is sensing them and varying its actions to maintain those consequences in relation to specific reference states. As disturbances vary their effects on the ultimate consequence, the organism varies its own effects in just the way that will cancel the effects of the disturbance. Since the consequence itself is being monitored, the source of the disturbance makes no difference and need not be sensed. (Sensing the cause of a disturbance creates "feed-forward," which is sometimes useful, but never essential.)

In brief, the old model says that organisms are organized to produce predetermined *actions*. Control theory says that organisms are organized to produce internally selected *perceptions*, which in many cases are perceptions of the same events that the external observer sees as the organism's behavior. The organism acts to bring under control, in relation to some reference state, the sensed perceptions.

The Properties of a Control System

The concept of control behavior as a process of error correction is helpful as a way to understand the general organization of control systems but is misleading with respect to most interesting kinds of control behavior. Given only what we have seen above, one might think that the brain selects or generates a reference signal, and that behavior is then produced by a control system that gradually brings its sensory signal to a match with the reference signal. There may be cases of complex and time-consuming actions that should be seen in this way, but I rather doubt it, and I will try to explain why.

Let's consider a case in which the driver might open the car door *part-way*, for instance, while backing out of the driveway and looking out of the door. This action involves setting a reference position for the door that is between 0 and 80 degrees, and *maintaining* the door in that position despite tilts of the car and gusts of wind.

To understand what *maintaining* implies, we begin with the door at its reference position, say 30 degrees. In this position, the door generates a sensory analog that exactly matches a reference signal generated elsewhere in the driver's brain.

If the door were actually at exactly that position, there would clearly be no error signal in the driver's brain. If there were no error signal, the driver's muscle systems would be driven neither to pull nor to push. If any disturbance then arose, there would be nothing to oppose its tendency to move the door, and the door would begin to swing to angles higher or lower than 30 degrees.

Suppose a disturbance swings the door more open. As the angle reaches 31 degrees, 1 degree of error appears. Suppose that for each degree of error, 100 grams of pull are generated by the muscles. Suppose further that the disturbance causing this error is equivalent to a force of 5 kilograms (5,000 grams) acting at the same place the driver's hand acts, but to open the door. We can see immediately that the door will continue to open until the error is large enough to produce 5,000 grams of pull by the muscles that will cancel the disturbing force and prevent any further opening. (We are neglecting inertial effects here.) How far beyond the reference position does the door have to be to generate a pull by the driver amounting to 5,000 grams? Fifty degrees, since each degree generates 100 grams of pull. The door will end up at an angle of 80 degrees, fully open.

The driver's *error sensitivity*, we can say, is 100 grams of effort per degree of error. Let us now suppose that this error sensitivity becomes 10 times as large, or 1,000 grams per degree. How far will the door now depart from its reference condition under a disturbance of 5,000 grams? Five degrees. It will swing to an angle of 30 + 5, or 35 degrees. A disturbance of the same size but tending to *close* the door would generate the opposite sense of error, producing a push; if the push error sensitivity were also 1,000 grams per degree (it need not be the same as the pull sensitivity), the door would swing to 25 degrees, 5 degrees less than the reference setting.

Continuing in this vein, if the error sensitivity becomes 10,000 grams per degree, a 5,000-gram disturbance would move the door only 0.5 degree from its reference position: at 100,000 grams per degree, the error would be only 0.05 degree. The higher the error sensitivity, the stiffer this control system would seem to another person pushing on the door—the less it would yield to a push.

If we say that an error of 1 degree is small enough to be unimportant with regard to any of the driver's purposes, it clearly ceases to matter what his error sensitivity is as long as it is greater than 5,000 grams per degree (and disturbances remain less than 5,000 grams). Furthermore, if the man's error sensitivity is 10 times that required minimum, there will be essentially no effect on the controlled variable if his error sensitivity *changes*—if it increases by any amount, or decreases by any factor smaller than 10. If the driver were required to hold the door against disturbances at an angle of 30 degrees for a very long time, his muscles would continually fatigue, reducing their sensitivity to neural signals and thus continually reducing the error sensitivity of this control system. But the door would remain very nearly at its reference condition until just before the end; only when the muscles have lost *most* of their sensitivity to neural stimulation will the door finally begin to yield appreciably to disturbances. After that point is reached, only a slightly greater disturbance or a few more minutes of loss of error sensitivity will result in an apparently sudden collapse of the ability to hold the door at its reference position. In truth, the decline in error sensitivity has been continual, but the properties of the control system have concealed that decline.

Given a control system with a high error sensitivity, what would happen if the door were nearly shut, but the reference signal in the driver's brain were equivalent to an angle of 80 degrees? The error would be 80 degrees. At an error sensitivity of 10,000 grams per degree, the driver would exert an initial force on the door of 800 kilograms, or about 1,760 pounds! In other words, the driver would be pushing as hard as possible, whatever that limit is. The door would fly violently open and probably ruin the hinges and the front fender.

There are many ad hoc solutions to this problem, involving variable control of speed and nonlinearities with just the right properties, but they are all complex in comparison to the one

I propose. I assume that in most human control behaviors, error sensitivity is very high, so high that under normal conditions there is essentially 0 error at all times. To take care of the deleterious effects of too much error, I simply assume that reference signals normally vary continuously, not in an on-off manner.

If reference signals normally change continuously instead of in jumps, and if ordinary rates of change are limited enough, the control system receiving a changing reference signal can keep its sensory analogue of the controlled variable matching the reference signal at all times. When the driver grasps the door and "opens" it, I assume that he starts with a reference signal corresponding to nearly closed, and smoothly increases that reference signal to the magnitude which corresponds to fully opened. The highly sensitive control system which is involved in this action keeps the sensory analogue of the door angle matching this smoothly increasing reference signal at all times, so the door opens in a manner that exactly reflects the smoothly increasing reference signal. Once the reference signal has reached the magnitude representing the intended final angle, it ceases to change, and the control system then maintains the angle essentially at the specified reference angle, countering all disturbances. The behavior we see does not represent a process of error correction, but changes in reference signals. The only time we would see the process of error correction itself would be right after a very large and sudden disturbance.

An Important Illusion

If no disturbances were acting, the driver could hold the door at a reference position of 45 degrees without using any effort, either push or pull. The sensory analog of the controlled variable would then exactly match the reference signal. Let us consider now the relationship between a pushing or pulling effort and a gradually increasing disturbance, assuming high error sensitivity in this control system and a constant reference signal.

As the disturbance begins to increase from 0, the error begins to increase, too, just enough to balance the push or pull against the disturbance. The system is always in equilibrium, because any less effort would allow the error to increase, creating more effort, and any more effort would decrease the error and the effort. If the error sensitivity is high enough, the error will never become significant in proportion to the reference setting, and the controlled variable will not change appreciably.

For all practical purposes, therefore, the door will remain at an angle of 45 degrees, and as disturbances come and go, efforts will appear in the driver's arm muscles to create pushes and pulls that are always equal and opposite to the disturbance. If there is a steady disturbance, the muscles will produce a steady force. If the disturbance is sudden and brief, the muscles will produce a sudden and brief opposing effort. The appearance will be exactly as if the disturbance acted as a stimulus to the man's nervous system, causing a pushing or pulling movement. This appearance, this illusion, is the basis for the model that has been used in all the life sciences since their beginnings. Even Descartes used it 340 years ago.

Organisms do not *react*. They *act*, and their actions always control some set of sensed variables inside or outside the organism. Every behavior that seems to be a simple reaction to a stimulus can be seen, on more careful examination, to control some variable; the apparent stimulus can be seen to involve a disturbance of that same variable. While there may be exceptions to this principle, I believe they are few and relatively unimportant.

A Hierarchy of Control

We now have the ingredients with which to build a more complex model, one capable of representing behavior organized out of many detailed behaviors at many levels. The examples in Table 1 can help to show how a hierarchical model would be built.

Consider the adjustment of the rear view mirror. The point of changing the angle of the mirror is not to achieve any particular angle, but to center the image of the car's rear window. To center this image, a number of actions must take place.

First, the driver must reach up and grasp the mirror. Then he must exert a twisting force to overcome the friction of the mirror mount without causing a sudden large movement. The force must then be varied in whatever way is required to make the image move toward the selected position.

Raising the hand to the mirror involves smoothly changing at least three reference signals, those defining the position of the hand in a three-dimensional space. These reference signals are altered from some beginning settings to some final settings. The efforts generated by the muscles keep the perceived coordinates of the hand (in some informal subjective coordinate system) matching those defined by the reference signals; three control systems, each controlling one dimension of this sensed position, would suffice.

We can now ask what is causing these reference signals to vary. The answer can be seen in common sense: they are varying to get the hand into the right relationship with the mirror to grasp it. The signals that are reference signals with respect to one set of control systems must indicate the *actions* of a higher-level system. This higher-level system senses the relationship of hand to mirror as it exists at each moment, compares this relationship to a reference-relationship, and converts the error into a shift in lower-level reference signals. That is all it can do—it cannot run any muscles directly.

The lower-level systems simply strive to keep their respective controlled variables matching the reference signal each is receiving from the higher-level system. If the higher-level system emits a fixed set of reference signals, the lower-level systems maintain the hand in a fixed position. If the higher-level system emits a changing reference signal, the lower-level systems, still keeping the sensed position matching this reference, produce a correspondingly changing hand position.

The lower-level systems, in other words, do not have to know anything about movements or relationships. They need only sense and control position, maintaining the position error as small as possible at all times. Any reasonable disturbance, such as the weight and inertia of the arm being moved, will be counteracted by the position control systems; the higher-level system that senses and controls a relationship never senses the effect of such disturbances, since those effects are cancelled at the lower level.

The higher-level system is sensing the same situation that the position-control systems are sensing; in fact, it must be receiving copies of the sensory signals indicating position, the same sensory signals being controlled by the lower-level systems. These signals contain information about hand position, which is one of the elements from which the sense of a relationship between the hand and something else is constructed. While it would be foolish to propose *too* detailed a model at this early stage of developments, it seems reasonable to think that higher-level systems in general perceive an environment that is made up of lower-level sensory signals, some of which are under control by lower-level systems and thus can be manipulated by adjusting reference signals that reach those lower-level systems. A higher-level sensory signal is derived from a set of lower-level sensory signals through continuous computing processes, so that the higher-level sensory signal presents a continuous report on the status of some aspect of the lower-order world, an aspect no lower-level system can sense.

I use the term *perception* to refer to signals derived from lower-level sensory signals. In this model, perception begins with the signals generated by sensory nerve-endings. Then there are successive stages of perception, each resulting from some continuous computing process that creates new signals depending on lower-level signals in regular, if complex, ways. The external world is not only represented in the brain; it is re-represented many times, each new level of representation being derived from those of lower level.

Along with each level of representation goes a level of control. At any given level there may be many control systems —hundreds or even thousands, although not all would be active in every situation. Each control system receives a reference signal from the next higher level, and makes its own perceptual signal match that reference signal by altering reference signals for many lower-level systems. The picture is actually more complex than that, because many control systems may act at the same time by adjusting reference signals for a common set of lower-level systems; each lower-level system then receives only a *net* reference signal, and obviously does not maintain its perception at the setting demanded by *any* higher-order system. It can be shown, however, that if the simultaneously active higher-level systems sense independent aspects of the set of lower-order perceptions, each can still control its own perception independently of what the others at the same level are doing, even though no system of that level can uniquely determine the reference signal for any lower-level system. For more details, consult a textbook on analog computing, and see the methods for solving simultaneous equations.

Under the old model, which also is hierarchical in organization, a higher-level system sends command signals to lower-level systems. These command signals are elaborated by the lower-level output processes into ever-more-detailed commands, until finally the lowest level is reached, and the commands are turned into muscle tensions. Those muscle tensions cause the movements which we call behaviors.

Under the control-system model, the command signals sent by a higher-level system to a lower one do not command any actions. They *specify perceptions*. The action taken by a lower-level control system will depend not just on this command signal but on the present state of the lower-level perceptual signal and on the amount and direction of any disturbances that may be acting to alter that perceptual signal. There need be no correlation between the motor activities and the command signal, for the command is not to do, but to sense.

However many levels there are in an actual human hierarchy, we can say two things about them with confidence. First, there is not an infinite number of levels; in fact I have been able to identify only ten, even by allowing myself considerable leeway. Second, the reference signals reaching the highest-level systems do not come from higher-level systems; by definition, there aren't any higher-level systems. The first observation does not constitute a problem, but the second does.

My answer to the problem of source for the highest level of reference signals is essentially to ignore it. If the system as a whole is organized as a hierarchy of control systems, the highest reference signals have to come from somewhere unless they are all identically 0. (Lack of a reference signal has the same effect as setting a reference signal to 0: the concerned control system then acts to reduce its perceptual signal to 0 and keep it there.) If the reference signals exist and come from somewhere that is not another level of control systems, they must come from some built-in source, or from previous experience via memory storage, or from somewhere else. It does not seem important right now to decide on the answer. Since a few reasonable answers are available, we can tentatively assume that magic is not involved.

To understand how the entire hierarchy of control works, one must try to imagine all levels in operation at the same time. Of necessity, the higher-level systems work on a slower time-scale than the lower ones; the greater complexity of perceptual processes introduces delays, and the requirement of maintaining stability (freedom from spontaneous self-sustained oscillations that destroy control) despite these delays demands even further slowing and smoothing of higher-level activities. As far as any higher-level system is concerned, the response-time of a lower-level system is 0; the perception being specified by the reference signal sent to the lower-order system varies *as* the reference signal setting is altered. Any lag in lower-level actions is necessarily shorter than the lag in a higher-level system. The higher-level system necessarily contains smoothing filters that average out any changes that occur over intervals comparable to its lag-time, so that lag-time appears to the system itself to be 0. Thus each level involves a "specious present," a definition of "an instant," which is longer than for a lower-level system. From the perspective of the system that establishes the relationship "hand grasping mirror," the transition from the initial relationship to that final one requires no time.

The lowest level of behavioral control systems (as opposed to biochemical control systems, which are not treated here) is the spinal reflex. A spinal reflex is actually a closed loop of cause and effect in which a muscle has effects on a sensory nerve via a short path through the environment, so short that most reflex loops can be traced entirely within the skin of an organism. The bulk of these first-level control systems involves the control of sensed muscle effort. The comparator for a spinal control system is a motor neurone in the spinal cord; this neurone receives a sensory signal from something affected by the muscle, and also a large set of convergent "command" signals from higher in the nervous system. The effect of an increased command signal is uniformly either an increase in the inhibiting effect of a feedback signal, or a decrease in the exciting effect of a feedback signal. In both cases, negative feedback results and the organizational requirements for control are satisfied. The signal leaving the spinal cord and going to a muscle, the so-called "final common pathway," is actually the error signal of the control system. There is one such control system for every voluntary muscle in the body.

While an upper boundary for this hierarchy is hard to define, the lower boundary is clear. As far as overt behavior is concerned, there are no systems lower than the spinal reflexes, the level-one systems. Level 0 is the outside world. All action is carried out by specifying settings for the reference signals entering level-one control systems. These reference signals do not tell those systems what to do; they tell them how much tension to sense.

A Possible Human Hierarchy

In the following pages, I will present 10 levels of perception (and by implication, control) that I think are reasonable guesses about our actual construction. Each level is defined as a *class* of perception. What I assume is that the brain, at birth or before, contains these levels in the form of specialized types of computing networks. Within one level there are initially no perceptual signals corresponding to the elements of adult experience; instead there are the materials from which can be constructed neural functions of particular types. Some levels, for example, may require short-term memory; if that proves to be so, the neural components required to construct short-term memory devices are present, although not yet connected in any useful way.

I assume that in an adult person, there are specific devices at each level, each device being physically connected so as to receive certain lower-level perceptual signals and to compute on a continuous basis the value of some function of those signals; the computed value is the next level of perceptual signal. We are not born with these specific devices; they are constructed on the basis of experience with lower-level signals created by contact with the environment into which the person is born. The construction does not, of course, involve the creation of any new neurones; it is done by creating new connections among the fixed set of neurones we inherit. *Classes* of perceptions are created simply because the brain is layered into classes of potential computing devices. A level-two perceptual function, I assume, cannot ever perform a level-three perceptual function simply because level two of the brain does not contain the required types of neurones or interconnecting pathways. We inherit, therefore, the potential of perceiving certain fixed classes of experience, although we do not inherit the ability to perceive any particular items at a given level. If this assumption is correct, we should find that all normal adults, in any culture, of any race, in any occupation, or with any degree of education, will experience a world made up of these same categories (although of course with highly varied examples within each category). A mathematician from Harvard and an African pygmy will have the same levels from lowest to highest; they will differ only in the way those levels have become organized: in content, not form.

There is, of course, another possibility. Perhaps if levels of perception exist, it is because reality is organized the same way and we have merely evolved to perceive what is really there. I would have to see a proof of "what is really there" before I could accept that, however.

Let us now look at these proposed levels.

1. *Intensity*. Level-one perceptions are the signals generated by sensory nerve endings. As the intensity of stimulation increases, the frequency of firing of sensory nerves increases. Whether this relationship be direct or transient, first-level perceptual signals themselves can vary only in one dimension: frequency. Thus each such signal can carry only magnitude information, there being no way for one signal to carry, in addition, identification of the source.

These signals can be experienced subjectively. They accompany every modality of experience. They are experienced simply as *intensity*, without regard to kind. It is perfectly possible to judge, using higher-level processes, the relative intensity of a sound and a light, a process which makes no use of any information about the sound or the light except *how much* of each is present.

Intensity signals that are under control are primarily kinesthetic; we call them *effort*, meaning not directed effort but simply amount of effort. When we pick up an object, we judge its weight largely by the effort-intensity signals created by picking it up. If a large effort-intensity signal results, we say the object is heavy.

Most intensity signals are not under significant control at level one. They pass on to level two where all intensity signals are received and, through neural computations typical of level two, are transformed into sensations.

2. *Sensations.* I define a sensation as a weighted combination of intensities. I mean by this term what is ordinarily meant: color, taste, sound, force, and all such elementary experiences. Most sensation-signals are not under direct control, but their hierarchical relationship to intensities is not hard to see. There can be no sensation of color if there is no light intensity experienced, but the reverse is not true. Thus color sensations depend on combinations of visual intensity signals from receptors in the retina, an assertion that should not startle anybody.

A common sensation is *warmth*. If a warm object (known to be warm, of course, by the way it feels) is moved over the skin, many different intensity signals come and go but there remains *one* unchanging sensation of warmth. The identification of the experience as warmth is independent of *which* warmth receptors contribute to the level-two perceptual signal. A sensation signal is appreciated as a *quality* of experience. We experience cold as different from warmth not necessarily because different receptors are involved, but perhaps because different weights are assigned to intensity signals and *different signals* result. It is hard to describe the difference between cold and warm sensations, but one is clearly *here* while the other is *there*, in some sort of experiential scale.

Kinesthetic sensations amount to directed vectors. Common level-two kinesthetic sensations are push, pull, twist, and squeeze, sensed not in relationship to anything else but simply as familiar sensations. These coordinated sets of efforts are organized in the brain stem. It is interesting that we can experience them so easily.

3. *Configuration.* While it is easy to experience the world as being totally filled with sensations, it is next to impossi-

ble to prevent these sensations from grouping themselves into recognizable associations, i.e., configurations. In the visual modality, we call some of these configurations *objects*. Others are simply arrangements of parts of a space against the background of other parts. The Gestalt concept of figure and ground applies at this level.

To control a configuration, it is necessary to alter some set of sensations. For the driver of the car to create the familiar configuration made by the image of the rear window in the rear view mirror, it is necessary to alter the detailed visual sensations involved in the image of the scene on his retinas (and, of course, to alter many kinesthetic sensations). On the other hand, it is clearly possible to alter all those same sensations without creating that particular visual configuration —and for that matter, without aiming at any particular configuration. If the visual sensations of edge, shading, and color were not present, no visual configuration could exist, since sensations are the elements of configurations. Thus from the standpoint of which kind of perception depends on which, and from the standpoint of which must be altered to control which, configurations are at a higher level in the hierarchy than sensations.

There are configurations in every sensory modality. They are the least units of experience that can be shown to depend on the existence of at least two different sensations, and the control of which depends on altering sensations.

The world of configurations is static, in the sense that maintaining a configuration at a fixed reference level results in a static situation. A system controlling a configuration (or one dimension of it) may be quite active as disturbances fluctuate or as its reference signal changes, but we judge the nature of a level of perception and control by the state of the perception when the reference level is fixed, not by the nature of the output actions of the system.

4. *Transitions.* When we experience a series of configurations that are carried from one to the next by some regular transformation, the result is a perception that is not a configuration but a sense of movement or change. The driver backing out of the driveway, looking back through the partly-opened door, controls this sense of movement by combined use of

the accelerator and brake, backing out at a *speed* that matches some reference-speed. A pianist executing a run controls the rate of change of pitch heard from the piano. A person who has just touched a hot frying pan is relieved to feel the pain *diminishing* by holding his finger under cold water. The stagehand at the control panel makes one spotlight fade at just the right speed and another increase in brightness, effecting a smooth transition from one color to another while holding brightness constant. A sportscaster slows the speed with which he is forming words to achieve a higher-level goal: finishing just as the second hand reaches the exact minute mark.

A controlled transition is a transition of lower-level perceptions at a particular rate; the reference signal for a transition-controlling system governs the speed of the transition without regard to beginning or end points. Obviously there can be no transitions if there are no lower-level perceptions, nor if the perceptions that exist remain the same. To control a transition, it is necessary to vary lower-level perceptions, yet lower-level perceptions can vary without any transition being controlled. Thus transitions are at a higher level than the perceptions which, by changing, give rise to the sense of transition.

I should mention a problem that is often raised in connection with fast transitions such as playing a run on a piano. It is possible for a level-four system to issue a series of configuration reference signals to the systems controlling the fingers faster than those systems can work. By *exaggerating* the reference signals, the level-four system can create large enough error signals in the third-level systems to result in the needed finger-movements, the reference signals being switched before the lower-order errors have come close to being corrected. (Actually, to correct them might entail breaking a finger or pulling a muscle loose.) Such extremely rapid transitions, therefore, are produced with very little lower-level control. The transition-controlling system can operate smoothly since the speed is not varying more rapidly than the control system's abilities can handle, but the lower-level systems are operating with very large error signals, and in fact are unable to maintain good control. We do not hear that lack of control, because of course we cannot perceive on a fast enough time scale to notice what the pianist cannot sense either.

5. *Sequence.* When the driver shifts from first gear into reverse, he grasps the gearshift, depresses the clutch, moves the gearshift forward, and releases the clutch slowly while gradually increasing the pressure on the accelerator pedal. Thus backing up the car involves control of perceptions at all levels through transitions—and at least at one more level. These individual actions will not result in backing the car unless they are performed in a particular sequence. Certain configurations must occur before others. Transitions must occur at the right speed, and they must occur in the right place in the sequence. Several lower-order control actions must occur simultaneously at certain places in the sequence.

There is thus a fifth level of perception and control, which I call the sequence level. A common name for familiar short sequences is *event*. (Note that the spoken word "event" is an event.) A level-five system perceives the lower-order world in terms of *what event is in progress*. Errors are seen as deviations from the "shape" of the event in space and time and are opposed by adjustments of reference signals for transition-controlling systems. (Those systems, in turn, produce the required changes in configuration, sensation, and intensity.) For a pianist to execute the run on a piano just mentioned, his level-five system need only perceive that this event is progressing properly, making small adjustments in the speed reference signal as required to maintain the event in its proper state (the state specified by a higher-order system).

I assume that for each event we have learned to recognize there is a separate perceptual function adapted to see the lower-level world in terms of that event. This is the general assumption; a perceptual function reports the state of the lower-level world, or that portion from which it receives copies of perceptual signals, as one single perception. Its output, the perceptual signal it generates, indicates by its magnitude the degree to which that one aspect of the lower-level world is present. Thus a given set of transitions and so on could easily result in perceptions of many different events at the same time —but generally only one of those perceptions will be large enough to matter. There may be a mutual inhibitory effect among perceptions of a given level, as occurs in the retina, so that the largest perceptual signal tends to suppress smaller ones at the same level, increasing the contrast between a strong impression of one sequence and weak impression of another, and similarly at any level.

Events are at a higher level in the hierarchy than any of the levels already discussed, for the same reasons. A uniformly spinning wheel exhibits transition without creating the sense of an event, but there can be no event unless there is some change of transition. To control an event requires altering transitions, but the reverse is not true.

6. *Relationships.* Shifting into reverse and beginning to back the car out of the driveway might be perceived as one event, in the case of an experienced driver. As this event proceeds, it must be maintained in proper relationship to many other events: looking back, opening the door, cars passing by, children riding bicycles, and so on. The term *relationship* is the one of importance at level six.

It is hard to explain what is meant by relationship without using the word relationship. It seems clear that there can be no *controllable* relationship unless the items related are potentially independent of each other. At least two items of experience must be involved. Relationships are defined by the way the behavior of one element of experience covaries with another element of experience. To say that a cup is on a saucer is to name a relationship we would not see if we perceived the cup and saucer as being carved from a single piece of material.

At all the lower levels of perception, one can find examples by casual examination of the world outside. Those examples —an event like the explosion of a firecracker, for instance —seem to be objective; they exist and need only be noticed. Relationships are not quite so easy to see as something having an independent existence. Of course it is easy to see that a cup is on a saucer, but it is hard to define just where that "on-ness" is. It isn't really anywhere.

This same problem actually exists at all the lower levels; we don't notice it because we don't normally ask questions like "where on the apple is its shape?" or "exactly when does a golf swing begin?" At the relationship level, it becomes more obvious that the observer is introducing something into the lower-level world of perception that would not otherwise be seen there.

If one says that a certain candy bar tastes sweeter than another, he asserts a relationship along the dimension of sweetness, the relationship called *greater than*. One perception of sweetness is more pronounced than another. This is clearly a judgment, a subjective perception. But at this level, the distinction between subjective and objective becomes uncertain. If a saccharimeter is used to determine degree of sweetness, an experimenter may well feel that an objective relationship has been found, because the meter indicated "2" for one candy bar and "3" for another. This way of achieving objectivity, however, only changes the perceptions; it does not do away with the need for someone to perceive a relationship, in this case the fact that "3" is greater than "2." There is no way to point to a meter and show where this relationship is. It is a perceptual interpretation.

We seem to be crossing a boundary at this level, but that is only a matter of custom and habit. A close enough look at any level would reveal the same problem. What we identify as familiar items of experience in an objective world dissolve into their lower-level elements on close inspection, and their existence becomes plainly a matter of how we choose or are organized to perceive.

One familiar kind of relationship is involved in what we term an *operation*, in the sense of an action that affects something else. By dialing a telephone, we can create the perception of a preselected voice—most of the time. By setting an alarm clock we can create the perception of a buzzer eight hours later. By pressing down on one end of a crowbar, we can make two pieces of lumber separate at the other end.

This kind of relationship perception enables us to control one perception as a means of controlling another. In most such cases, we do not perceive *how* this comes about; we simply know that if one perception is brought to a particular reference state, a related change will occur in another perception. We perceive cause and effect, but not the processes that bring it about. Nevertheless, we can control such relationships, since we have the ability to vary one or more of the related elements to counteract changes that we cannot control in the rest of the elements. A cat chasing a mouse cannot control the movements of the mouse, but by altering its own motions, it can control the relationship between itself and the mouse. Even cats have some level-six systems.

Relationships perceived at this level can vary from the elementary ("toward") to the complex ("mother"), from the formal ("exclusive or") to the informal ("nicer"). Most relationships that we control are probably perceived but not named; the relationship, for example, between oneself and a partially-opened door through which one is squeezing, or between a fork and a pile of scrambled eggs one is eating. To perceive and control relationships one does not have to talk about them; talking is a different sort of activity. Most of what we perceive and control at all levels is nonverbal; it's just that some of us tend to limit our conscious attention to the verbalizations.

Relationships are of higher level than events, by the same criteria we have been using. It may be a good exercise for the reader to apply them explicitly.

7. *Categories.* This level did not appear in my 1973 book, and it may not survive much beyond this appearance. The only reason for its introduction on a trial basis is to account for the transition between what seems to be direct, silent control of perceptions to a mode of control involving symbolic processes (level 8).

A category or class is truly a disembodied entity. If I perceive a familiar shape, I might call it "Fido," or I might call it "a dog." If I call it "Fido" I mean to point to a particular configuration, that one right over there, my dog. If I call it "a dog," however, I am really not pointing to that dog at all. I am indicating a *class* of perceptions of which that particular one is only an example. If I say, "A dog will eat dog food," I do not mean that *my* dog, that one there, would eat dog food if I set it down in front of him—my dog might be full at the moment. I mean that the class of items called dogs performs the class of activities called eating relative to the class of items called dog food. I am speaking in classes, not specifics.

To know what class name to apply in any given case, it is necessary first to be able to distinguish among classes, perceptually. Should I call that relationship "racing" or "fleeing"? I cannot pick the name until I have made a perceptual identification. I make the identification by examining an array of relationships, and if the relationships make a familiar pattern, I "recognize" the category, after which I can come up with its name.

It is often said that classes or categories are established by looking for something which different items have in common. I think that is backwards. What we really do is to establish first the idea of a class, say items with "one broken leg," and then look to see if the items at hand can be perceived as members of that class. By working back and forth between lower levels and the category level, we often can come up with a familiar category that can then be exemplified by all the items being examined. The general concept of "one broken leg" has to be organized as a mode of perception before we can see if anything, even a single item, belongs to that category.

Many useful categories are formed so that their lower-level examples entail clearly recognizable perceptions of the same kind. This is definitely not true of all categories, however, which is why I am rejecting the usual concept of common elements as the determinant. Consider the category of things that are "mine," or things that are "expensive," or things that are "not here." A category is basically an *arbitrary* way of grouping items of experience; we retain those that prove useful.

This means that we can quite easily treat items as members of the same category even though we see nothing in common among them. The particular example I am thinking of, and the main reason initially for considering this as a level of perception, is the category which contains a set of perceptions (dog 1, dog 2, dog 3, dog 4 ...) and one or more perceptions of a totally different kind ("dog," a spoken or written word). Because we can form *arbitrary* categories, we can symbolize. The perception used as a symbol becomes just as good an example of a category as the other perceptions that have been recognized as examples.

A symbol is merely a perception used in a particular way, as a tag for a class. The perception of a particular class can then be evoked as easily by that tag as by any of the other perceptions that are also perceived as examples of that class. The process also works the other way; if I ask, "Have you ever had one of those?" and point to Fido, you will understand that I am asking if you have ever had "a dog."

At this level maps and territories begin to get confused. If one too regularly confines attention to perceptions of this level or higher, he may forget that treating different items in terms of the category to which they belong is ignoring the differences making them unique. Those differences become evident only if all levels of perception are equally available to awareness.

To speak of "control" of a category may seem strange if one thinks of it in the same way as control of position. Remember that to control something is basically to do what is necessary to create a specified perception of that something. When we come to the category level, the states of perceptions tend to become black-and-white; either this category is exemplified, or it is not. This is a dog, or it is not a dog. So to control a perception of category may require nothing more than to bring about one example of it in lower-order perception, perhaps just by looking in the right direction, or perhaps by going through lower-order actions that will reveal critical perceptions resulting in perceiving one category rather than another. ("I am looking for a nice dog for my nephew.")

Finally, it may seem that establishing a category-perception must involve a very complex and mysterious computing process. I don't think so; I think the process is almost trivially simple. All that is necessary is to "or" together all the lower-order perceptual signals that are considered members of the same category. The perceptual signal indicating presence of the category is then created if any input is present. In fact this process is so simple that I have doubts about treating it as a separate level of perception, despite its importance. The logical "or," after all, is just another relationship. It may be that categories represent no more than one of the things a relationship level can perceive.

8. *Programs.* The reason I want category perceptions to be present, whether generated by a special level or not, is that the eighth level seems to operate in terms of symbols and not so interestingly in terms of direct lower-level perceptions. At this level, we have what are called *contingencies.* If one relationship is contingent on another—if a grapefruit will fit into a jar only if its diameter is smaller than that of the jar's mouth—we can establish the contingent relationship if the other it de-

pends on is also present.

To perceive in terms of contingencies, one must understand a branching network of possibilities. *If* condition A holds, take branch 1; otherwise take branch 2. Some kind of *test* is implied as part of the perceptual process.

I don't want to try too hard to make this level fit the pattern of the others. Perhaps it is best merely to say that this level works the way a computer program works and not worry too much about how perception, comparison, reference signals, and error signals get into the act. I think that there are control systems at this level, but that they are constructed as a computer program is constructed, not as a servomechanism is wired.

For example, perceptual processes can be constructed to rely quite specifically on rational computing processes. I see a green Ford in the street. It has Missouri license plates. There is a steaming puddle between its front wheels, and its grill is mangled. Aha, I think, somebody from out of town has had an accident. That statement amounts to a perception of the situation constructed out of lower-level perceptions by a process that must be called reasoning. That perception appears as a string of symbols. I may then realize that it is my civic duty to report the accident (civic duty representing a reference signal from a higher level) and decide to call the police. That decision, of course, specifies not a particular action but a *class* of actions: my eighth-level system has selected a reference signal for my seventh-level systems. It is then up to my seventh-level systems to find specific relationships and events that will provide a perception of the class, "calling the police." If there is a policeman passing by I can call out to him; otherwise I might use the telephone.

Here is another example. An engineer has the responsibility for maintaining the density of a batch of paint at a constant level. He records the volume of paint in a container and the weight of the paint plus container. Subtracting the weight of the container, he gets the weight of the paint, and by dividing the weight by 32 times the volume he obtains the density as a number. All of this is done through manipulation of symbols, on paper or in his head. He then compares that number with a reference-number, subtracts to obtain the difference, and, depending on whether the difference is positive or negative, adds solvent or pigment to the container. He repeats this process until the density is what it should be.

This is a perfectly good control system; it simply works in terms of computations carried out in symbols instead of lower-level perceptions. People can obviously do this sort of thing; therefore they obviously need a level of organization capable of performing the necessary operations. I can't think of a better reason for putting this level into the model. Miller, Galanter, and Pribram (1960) proposed a whole model of human organization based on this sort of programmed control concept.

Operations of this sort using symbols have long been known to depend on a few basic processes: logical operations and tests. Digital computers imitate the human ability to carry out such processes, just as servomechanisms imitate lower-level human control actions. As in the case of the servomechanism, building workable digital computers has informed us of the operations needed to carry out the processes human beings perform naturally—perhaps not the *only* way such processes could be carried out, but certainly *one* way, which is better than not knowing any. Knowing one arrangement of components that can imitate aspects of human reasoning, we can be confident that reasoning is not carried out by little men in our heads or by magic; we can accept it as part of our understanding of human nature without defying physics or any other set of principles we have reason to accept.

There is one problem which experience with digital computers has revealed very clearly. Any machine that can be programmed can be programmed in an extremely large number of different ways; for a human brain, or even just one of many levels of organization in it, that number is for all practical purposes infinite. Among the infinity of programs that might run in the human eighth level of organization are *hierarchically organized* programs. Furthermore, any system which we are capable of understanding can be *simulated* in a computer: the physical variables can be represented by variable numbers, and the relationships among the physical variables can be replaced by computations. Thus the entire hierarchy we have been looking at could be repeated as a program in our eighth-order systems. Indeed, isn't that what we have been doing here?

Worse than that: not only this hierarchy, but any other could be simulated, including a hierarchy with 200 levels, or 2,000, or 2. These hierarchies, if debugged, could run quite successfully, and the person programmed this way at the eighth level could act as if his brain were organized, wired, as the program is organized, at least as far as his symbol-manipulating activities were concerned.

I call this a problem (it is also a tremendous advantage for us) because of the confusion it generates among those who try to model the organization of human beings. Most of the models I know about are not really models of the brain like the one we have been going through; they are models of programs that can run at the eighth level of a brain. Semantic network analysis is that sort of thing. Miller, Galanter, and Pribram's TOTE unit is that sort of thing. Most abstract mathematical models, even those under investigation by some pretty good and smart friends of mine, are that sort of thing. Studies of artificial intelligence, indeed all systems that depend primarily on qualitative and verbal reasoning, are also.

Once we know that programs can run in the brain, there is little point in getting involved with the question of *what* programs can run. That is an interesting question, but here we are trying to identify levels of organization that *must* be there, not those that represent only happenstance culturally supplied information content. If there is a level of the brain that can be programmed by stored information, that level can behave like any conceivable organization: hierarchy, heterarchy, stimulus-response robot, or Turing Machine. It can be programmed to behave rationally or irrationally. It can be programmed as a positive feedback system that destroys itself as soon as it acts. None of those details tell us more than we knew when we realized that programs can run in a brain.

When we get to the subject of development, by the way, I think it will be clear that studying programs can be quite important when the point is to investigate one person's capabilities.

9. *Principles*. We now go beyond the levels where I think there is a clear justification for assumptions. If there is a level higher than the program level, it would be the level that perceives something exemplified by many different programs,

and specifies objectives for the program level to accomplish. Naturally, since the program level can be and probably is organized hierarchically, it is not easy to specify a higher level that is not just another level of programming, but I think it can be done.

The sort of perceptual level we are after is the kind one would use in evaluating a program, or in choosing one program over another. This program, one might say, is inefficient; that one is clumsy; the next is elegant; the last violates privacy. Considerations come into play that have nothing to do with any particular program.

I have chosen the term *principle* to characterize perceptions of the ninth level. There must be a better word, but this one will have to do for now. The meaning I intend is in the sense of *principle of operation*, as would be perceived when one looks at a program and says, "I know that one—it calculates square roots by *successive approximation*." Successive approximation is a principle. It is not any particular program, although particular programs which are examples of it can certainly be devised. Once one has understood the strategy of successive approximation, one can recognize it in any disguise. As far as I know, nobody has been able to write a program which, by itself, could obey the command "Use successive approximation!" Programs just can't generalize in that way. Programmers, however, can.

Another way to think of the ninth level might be this. Suppose there is a square root (or a shopping-for-potatoes) program in operation. At the program level, the activities would involve manipulating symbols according to preprogrammed rules and applying contingency tests at appropriate places. Nowhere in that program, however, is anything capable of characterizing the program that is running; the program is running, not characterizing. For a person to *describe* the program, it would be necessary for a point of view to exist from which the program could be seen in operation—and that would have to be outside the program.

However many levels of hierarchically organized programming are running, there is a highest level. In order to characterize that level, a person needs a point of view that is higher still: that is, higher than the program level itself. That higher level selects first this executive program and then that one, to bring about perceptions that may include programs themselves as elements, but which are not of their nature. This higher level would decide, for example, "I'm never going to solve this problem by thinking; it's time to try something at random."

In that unfinished state, we leave level nine.

10. *System concepts.* We now venture onto shaky ground indeed. I think that one more level is needed to take care of a perception that still seems missing. Consider not the word "I," but that entity to which the word refers, for you. The referent of this word consists of essentially all we have been through, *plus* something else, which I call a system concept. If I say the word "physics" to a physicist, he knows what it means: a system of principles, procedures, relationships, events, and so on to the bottom. But these elements all comprise a *system*, not an assortment. It hangs together, creates the impression of one organized entity. We have many words—not many in comparison to other kinds, but quite a few—which point to an experience we all recognize; the concept of a person, a country, a company, a self, a family. We can clearly tell the difference between different system concepts; each one creates a context in which everything else happens. This context is almost tangible, and sometimes is treated as if it were more tangible than its elements. (Think of a baseball team after a complete change of players, coaches, managers, owners, and city.) When the system concept in question is one's self, one will go to considerable lengths to preserve its integrity; an error signal at that level is taken very seriously. I'm not quite sure what good it does us to have and preserve reference states for system concepts, but I think we clearly do. And this is as far as my thinking has so far progressed.

Table 2 is a summary of the ten levels, with examples drawn from a number of different contexts.

A final remark, before we go on to look at applications, is in order. It may seem that all the levels higher than the first would be invisible internal processes, akin to intervening variables, and thus not directly testable. In fact all these levels can be seen in action from outside the organism. Each deals with a different aspect of the perceived world, and to the extent that

Table 2. Some possible levels in a hierarchy of control systems.

Level	General Term	Music	Physics (inanimate world)	Motor Activities	Inner Experience	Verbal Experience (auditory)	Verbal Experience (visual)
10	System	Performance	System	Occupation	Being		Communication
9	Principle	Interpretation	Principle	Style	Attitude		Deep structure
8	Program	Execution	Process	Technique	Thought		Grammar
7	Category	Opus	Phenomenon	Task	Role		Semantics
6	Relationship	Orchestration	Interaction	Operation	Interaction		Syntax
5	Event	Phrase	Event	Pattern	Action		Word/phrase
4	Transition	Dynamics	Path	Motion	Change	Consonant	Direction
3	Configuration	Chord	State	Posture	Emotion	Phoneme	Letter/word
2	Sensation	Pitch	Property	Effort	Feeling	Pitch/hiss	Mark
1	Intensity	Loudness	Magnitude	Tension	Intensity	Loudness	Brightness

all human beings perceive in about the same manner, one human being can see variables of any level being controlled just by paying attention to the right aspects of another's actions. As we shall see, all proposals about control systems in this model can be tested by acceptable experimental methods.

The ten levels we have just seen are one concept of an adult's organization, or the inherited framework within which that organization comes into being. It matters very little whether those ten levels are correct. The point of developing them has not been to construct an immense hypothesis to test in one Grand Master Experiment, but to survey the range of organizations in a human being that need to be accounted for and to get a feel for how all human activity might be handled in one consistent theory.

There are those, no doubt, who feel that efforts like these are a waste of time, but I do not agree. Psychology has, in the past, used models so elementary in their structure that it is an insult to be told they represent human beings. The hierarchical control system model represented by these ten levels is not too complex; it is still too simple. Its main advantage is that it deals with at least a respectable range of human activities without the use of a Procrustean bed, and does so in a way we have a hope of understanding.

I have laid out this hierarchy in so much detail to demonstrate an approach. Everything that I have been able to catch myself doing while constructing and using this theory is in that model. I have claimed no privileged point of view. It may well be that I have not named the levels correctly, or that under different circumstances different levels would be observed, but the existence of hierarchical control levels in which one system acts by determining the reference signals for others seems to me a sound basic hypothesis. The need to account for organized activities from spinal reflexes to rational thought and beyond seems far clearer to me now than before I had constructed these levels; I hope that is true for the reader, too.

As I said in the beginning, thinking of human organization in terms of these levels must have a strong influence on what we consider human development to be. Having seen one coherent set of levels that encompasses most aspects of human activities, perhaps it will be possible for others now to begin asking questions that will reveal better definitions or aspects of organization I have missed. There can be no better subject than the study of human development for this purpose. Let us now see how this kind of study could proceed in the light of a hierarchical control concept.

The Discovery of Control Organizations

There is, fortunately, a systematic way to test any hypothesis about human control behavior. Indeed, if that were not true, this whole theory *would* be an untestable fantasy. In a way, I have applied the method to be described here all during the development of this model, although not with the formality that must eventually be demanded. The essence of the method is a procedure to test the hypothesis that some variable is under control. With this method, one can feel much freer than otherwise to propose possible control organizations, since it will ruthlessly weed out wrong guesses.

I call this method a test for the controlled variable. It functions as follows.

Suppose that for any reason (or for no reason) one decides to test the hypothesis that variable A is controlled by a control system. Available are other physical variables B, C, D, and so on, each influencing the state of A. The nature of these influences can be found by inspection of or experiment with the variables, in the absence of any potential control system to interfere with the inspection.

The test involves nothing more than applying disturbances to the selected variable by altering other physical variables known to affect it, and verifying that the definition of control is satisfied. If it is satisfied, every influence that should affect the state of the controlled variable is met by some equal and opposite reaction from a control system; the result is that the proposed variable is stabilized against disturbances. It is not necessary, initially, to know where the control system is or what its mode of action is. The first step is always to try disturbing a proposed variable, and to see whether the resulting changes in the variable can be accounted for strictly in terms of observable physical phenomena, including the disturbance. If they can, there is no control system.

If a variable is affected far less by a disturbance than expected, something else is acting to oppose the effects of the disturbance. To be sure that this something else is a control system, one must identify the *means* of control, and also show that the variable has to be *sensed* by the system applying the means, for control to be maintained.

A Thought Experiment

Let's try this method in a thought experiment that should be familiar enough to most readers to make the conclusions seem realistic. We will occupy the passenger seat while our driver drives the car down a straight level unencumbered freeway on a windless day. One obvious hypothesis is that the driver is controlling the position of the car relative to the road.

From previous experience with cars, we know that even a small steady deflection of the steering wheel will cause a car travelling 55 miles per hour to move quickly out of its lane. Thus it would be easy to insert a disturbance; we quietly take hold of the steering wheel and exert a small force to turn it. The result is *not* the expected veering of the car, so we know that the position of the car in its lane may be related to a controlled variable.

The disturbance is clearly being counteracted by a force from the driver's arm muscles; it is not hard to discover the means being used. We shall now see why it is necessary also to check the perception that is involved.

Suppose we temporarily alter our hypothesis, because we can now see that *steering wheel position* is also being held constant and is much more directly involved in our disturbance and the driver's reaction to it. It would be hard to interrupt the driver's ability to sense steering wheel position since he can *feel* it as well as see it as long as he is holding the wheel. So let us provide a different means of control that does not involve the driver's holding onto the wheel: *we* will move the wheel according to verbal directions from the driver. Each time he says "left" we turn the wheel a little further to the left; each time he says "right," a little further to the right.

The driver won't be able to steer quite as precisely under these conditions, but we will find that he can still keep the car well within its lane with a little practice. Now, if we persuade the driver to allow a cardboard shield to be placed where it blocks his view of the steering wheel, we can cut off his means of sensing the wheel position without preventing him from *affecting* the wheel position.

With the shield in place, we test the driver's ability to resist disturbances of the steering wheel again. If we do not get too violent or make very rapid changes in the disturbance, we will find that the driver countermands every disturbance we apply independently of him. The result is that the steering wheel is never permitted to stray very much from its original position.

The test has thus been failed. Steering-wheel position is *not* a controlled variable, because the driver continues to act much as he did before even without being able to sense steering wheel position.

We could quickly establish the real nature of the controlled variable by returning to the original conditions, and this time blocking the driver's view of the road. As soon as we prevent the driver from seeing the relationship of the car to the road, control is lost. The driver may try to resist our disturbances of the wheel—he probably will—but he will not be able to cancel disturbances exactly enough to keep the car on the road. He will have switched, in fact, to controlling the sensed position of the wheel, as we could verify by adding the cardboard shield again and going to the verbal mode. In that case, control of wheel position would be lost, too, as anyone would predict who is clearly visualizing the situation.

If we were doing this in a driver-training simulator, we could narrow the definition of the controlled variable even further. In a good simulator, a computer translates the speed of the car and the steering wheel angle into appropriate changes in the visual display. By tampering with the computer, we could disturb the visual display relative to the computed position of the car, displacing the display by some small amount to the right or left of where it should be. Now we find that the computed position of the car varies every time we displace the display; what stays constant is the display, not the position of the car. The picture that the driver sees in the windshield is, or contains, the controlled variable. By fuzzing out various aspects of the display, we could eventually discover just what aspects of that picture are being sensed, and hence just what variable is under control.

Something similar could be done with speed. By surreptitiously pressing down on the accelerator, we could test the idea that the driver is controlling (a) accelerator pressure, (b) rate of change of the scene in the windshield, or (c) the speedometer reading. The reader can, of course, now design the necessary experimental details.

Exploring the Hierarchy

The above test provides a way for identifying controlled variables for a system in which the reference signal remains constant for the duration of the experiment. Once the principle is understood, this test can be carried out quite rapidly, so the reference signal need not remain constant long, but the demand for a constant reference signal does limit the application of the test in a hierarchy of systems. The reason is that in order to maintain a higher-level variable constant, a control system acts by *continually altering* lower-level reference settings. We need some way of identifying control systems of lower level than the principal one under investigation.

It is sometimes possible to identify a few lower-level organizations in natural situations, because lower-level systems always operate on a faster timescale than higher-level systems. In the case of the driver steering a straight course, we might well suspect that there is a lower-level system for controlling the position of the steering wheel, as part of the means for controlling the car-to-road relationship. Applying brief disturbances to the steering wheel, we could verify that resistive forces develop *before* any appreciable change in car position can occur. There is an initial immediate resistance, which then becomes larger, after delay, when the car actually moves visibly in its lane. In effect, we are seeing the lower-level system operating with a constant reference signal for a short time, before the higher-level system senses an error in its variable and alters the lower-level reference signal.

This "nesting" of reaction times provides one way, probably best suited to the laboratory, for sorting control actions by level. It is not, of course, absolute; I have found, for example, that separation of levels is not good unless the control behavior is very well practiced. Clear evidence of control does not appear until learning is essentially complete.

A more general method can often give much more direct evidence of levels of control. Each level in the hierarchy deals with a different class of variable. Furthermore, at every level there are many perceptions that are *not* under control; think of the previous example of a cat controlling its relationship to a mouse without being able to control the sensed position of the mouse.

It is therefore often possible to apply disturbances which affect a system of one level without affecting systems of lower level. In the case of the driver, we have been applying disturbances that affect a possible wheel-position control system as well as the car-position system that is supposedly using that wheel-control system; it is awkward and in many cases impossible to separate the two levels using this sort of disturbance. But there are other kinds. Suppose, for example, that we had a way of keeping track of the crosswind component of any wind that was blowing, and had tables for converting that component into an effective disturbance of car position. The wind is not going to disturb the steering wheel directly; hence even if there is a wheel-control system acting, it will not be disturbed. A sudden gust will not result in an instantly appearing muscle force; only as the car begins to drift sideways will the driver's arms begin to turn the wheel to oppose the effect of the wind.

The converse can sometimes be done. If we can apply *two* disturbances, one to the steering wheel and another (applied to the car directly) continuously adjusted to keep the car's position from changing, we can prevent the higher-level system from seeing any error, and thus prevent it from altering the lower-level reference signal. In effect, we take over the higher-level control action, and are able to see the lower-level system acting with a constant reference signal. We cancel higher-level disturbances before the system we are investigating can sense their effects. We can then disturb the variable for the lower-level system without causing the higher-level error that would usually also result.

How and Why

In most ordinary behavior, a person is engaged in control activity at some particular level that is more highly visible than any other level. This most visible level is the lowest level at which the reference condition for a controlled variable is being held, for the time, constant. A person standing at a bathtub with one hand under the water faucet while the water runs is waiting, perhaps, to feel a sufficient level of warmth before stoppering the tub to fill it. For the moment, we are seeing a simple sensation-control system in operation; the hand on the cold-water handle turns the handle until a steady level of warmth is felt. What is going at a higher level is a sequence called "filling the tub," or "taking a bath." For the time being, however, all reference signals above a certain level are fixed, until the current element of the sequence appears in perception.

If we pick an example of behavior at random, we have no immediate way of saying what level of organization is visible. By asking *how* and *why*, however, we can begin to get an idea of what level we are looking at, at least in relative terms.

To ask *how* a given perception is being controlled is to ask about the *means* of control. The means, for a system above the lowest level, is always manipulation of reference signals for lower-level systems, so asking *how* is the same as asking what lower-level variables are being controlled to control the higher-level variable. *How* does the person control the sensation of warmth? By exerting muscle efforts, the hand already being properly positioned on the handle. Once higher-order systems have taken care of that positioning and have established a firm grasp, a simple two-level control system can control the warmth.

If we had started with the level of "taking a bath," asking *how* this event could be created would lead us, sooner or later, to the act of controlling warmth. Asking *how* takes us downward in the hierarchy.

Asking *why*, on the other hand, takes us upward. The meaning of "why" is not quite as unambiguous as the meaning of "how," but we will take it to mean, "for what purpose?" For what purpose is the person adjusting the warmth of the water? To fill the tub—and at least as one consideration among many, to control the transition of temperature that is felt as one first steps in. Ultimately, the purpose is to take a bath. More ultimately, to get clean. Still more ultimately (this being a hierarchy of many levels), to take care of one element of a relationship with other people.

When we ask *why* a given reference condition is being established, we automatically view that reference condition as an adjustable part of controlling some higher-level perception. In the spirit of the kind of modelling we have been doing here, one would always try to answer this *why* question in terms of the *least* upward step possible; one would not answer the question as to why a person is controlling the warmth of bathwater by saying "in order to be a civilized person," although for a given person that might be a system concept that is in fact under control by means (among other means) of taking a bath. The question *why*, as well as *how*, must be considered as unique to every individual. People take baths for all kinds of reasons. If we can map an individual's control organization, why try to characterize that person in fuzzy universal terms?

By exploring organization through asking *how* and *why*, we find details that do not commit us to any scheme of levels; we take the organizations we find. The test for the controlled variable then permits us to test hypotheses suggested by this less formal means of exploration. A fairly well defined methodology seems possible within the framework of this control model, a methodology that is quite independent of the particular levels I have proposed and which can be applied to any kind of behavior in any circumstances where the necessary observations can be made.

Let us close this chapter by looking at one way a program of research in human development might be constructed, using the tenets of this theory. In this closing section, we will finally discuss the one main field ignored until now. We now have a model of *performance* for a mature organism and ways of exploring organization. The question is now, how does that organization come about? We have to consider *learning*. I shall, however, consider it only obliquely.

Research in Human Development

I think it is clear that no person is born with a control system that controls the perception of being civilized, much less one which entails taking baths to accomplish that goal. There is no need to assume that we are born with any behavioral control systems at all, save perhaps a few constructed in the last four or five months in the womb and being centered around control of efforts.

In my 1973 book I proposed a model for learning in which a process called *reorganization* was driven by errors that signified some deviation within the organism itself: *intrinsic* error signals. The controlled variables for this reorganizing system were measures of the state of the organism; the reference conditions were inherited. The action of this system involved *altering connections* in the brain. This connection-altering process resulted in the hierarchy of control systems in an adult, and in the gradual creation of these control systems during development. The only preorganization I assumed was in the kinds of components available at various levels of the brain from which to construct the parts of control systems. The reorganizing system was not "intelligent." It did not learn from experience, nor did it seek any particular organization of the brain. All we call intelligence resided in the hierarchy of perception and control that results from reorganization.

This reorganizing system, clearly, has to be defined as what it does, not what it is. My proposals were an attempt to guess what it does. When we are talking about research in human development, however, we do not need to guess about that. We can find out by applying the principles of control theory to characterize human organization at various stages in its development and, on the basis of changes we see in that organization, we can specify what any reorganizing system must do.

My suggestions for a cybernetic research program in human development, therefore, will have nothing to say about learning theory itself. Just as most astronomers ceased to speculate about the surface of the moon once it was evident that men would soon be walking on it, I feel little urge to speculate about the process of reorganization, knowing that we will soon know more about it than we do now.

I see a program of research in human development as a series of stages we go through to build a new understanding of human nature in a systematic way. We first gather basic data in a more or less naturalistic way, data which traditional methods have not provided and could not provide. Then we try to classify this data into a taxonomy of levels—or else conclude that levels are not significant features of organization. Next, we can begin to look for patterns relating to the development of organization from very little to maximum. And ultimately, I hope we can devise efficient and accurate ways for mapping the organization of an individual from his most elementary reflexes to his overall properties. Out of that will come immense benefits. For the first time, we will be able to say just what is wrong with a person who is having troubles. We will be able to devise psychotherapeutic methods that go directly to the difficulty instead of taking years to find the trouble by a random search, or solving the problem by destroying capacities of the brain. We should be able to improve the methods of education by orders of magnitude. We should be able to match a whole person to a whole job, on the basis of specific knowledge rather than statistical comparisons with masses of other people. I have, in short, some definite and hopeful expectations about what we could do with control theory. The following are some ideas about how to get from here to there.

Gathering Data

The principal kind of data that is needed about any human being at any stage of development concerns *what variables that person can control*. If no human being controls any variables—if all hypotheses fail the test for the controlled variable—we can save ourselves a lot of trouble by abandoning the project there.

The initial effort, it seems to me, must be to accumulate experience with control behavior. I envision not just a vast effort at reinterpretation, but a systematic testing application for the controlled variable. The only data about human control behavior that will be of any use will be catalogues of controlled variables which have passed every application of the test for the controlled variable, conducted with as much rigor as possible. It does not matter if this data gathering is systematic; it may well be best to take examples at random, without regard to any proposed hierarchy or any other preconceived notions. Experimental verification of proposed controlled variables should be done with human beings of all ages, occupations, races, cultures, and so on. The greater the range of hypotheses tested, the better.

Classifying the Data

Once a sufficient amount of information is available in terms of variables proven to be controllable by human beings, the next step is to sort this information into meaningful categories. One obvious dimension of sorting is by age; at what age range does each variable become controllable? Another is by type, if that proves possible. I would be interested, of course, in types which can be shown to stand in hierarchical relationship, although any data is good data even if it does not support my organizational prejudices.

This phase will call for new kinds of data gathering, aimed at sharpening methods for distinguishing levels of organization, or for otherwise mapping relationships among control activities. There will be plenty of room for ingenuity.

Characterizing Development

There are reasons to think that in acquiring any new control organization, a person is likely to develop the parts of that system in a particular order: (1) perception: the variable to be controlled must exist as a neural analogue; (2) recording: the possible states of that variable must be experienced and remembered; (3) selection: one previous state of the variable must be selected as a reference signal; (4) comparison: the error between the actual and intended states of the variable must be judged; (5) action: the error must be converted into those changes of existing lower-level reference signals that will correct the error; and (6) practice: this entire series must be iterated over and over to refine each element of the control system so that it functions under all conditions without instability.

That is, of course, a proposal about how development of a new control system might occur. The objective of an experimental program would be to see if this sequence or any other is sufficiently universal to apply in general. If there is a preferred sequence for constructing a new control system, the implications for education or for learning skills are obvious.

More generally, the data existing by the time this phase of research became important would contain information about transitions from one organization to another. That data would contain the character of the reorganizing processes that underlie development. In traditional psychology, experiments are almost entirely concerned with learning, yet ways to characterize what is learned are as primitive as guesswork. By concentrating first on characterizing performance and organization, rather than changes in arbitrarily chosen measures, we should arrive at a learning theory that has more than statistical validity.

My hunch is that through a research program like this we will find that levels of control do exist, and that the development of a human being from fetus to adult centers around the acquisition of higher and higher levels of control and wider and wider varieties of control systems at each level. I have tried to envision a program which would not inevitably lead to that result, but which, should it uphold this hunch, would be believable on its own merits. I hope that there are others who see a challenge in this theory and this proposal and who will commence the work at hand.

References

Buckley, W. 1968. *Modern Systems Research for the Behavioral Scientist.* Chapter 5: Cybernetics, purpose, self-regulation, and self-direction. Debate with Rosenblueth, Wiener, and Bigelow on one side, and Taylor on the other in a series of papers. Chicago: Aldine. Pp. 219-242.

Maxwell, J.C. 1965. On Governors. In *The Collected Papers of James Clerk Maxwell*, W.D. Niven, ed. New York: Dover. Vol. 2, pp. 105-120.

Mayr, O. 1970. *The Origins of Feedback Control.* Cambridge,

Mass.: M.I.T. Press.
Miller, G.A., Galanter, E., and Pribram, K. 1960. *Plans and the Structure of Behavior.* New York: Holt, Rinehart, and Winston.
Powers, W.T. 1973. *Behavior: The Control of Perception.* Chicago: Aldine.
Watson, J. 1919. *Psychology from the Standpoint of a Behaviorist.* Philadelphia and London: J.B. Lippincott Co. P. 13.
Wiener, N. 1948. *Cybernetics: Control and Communication in the Animal and the Machine.* New York: Wiley.

Recommended Readings

Hayek, F. 1952. *The Sensory Order.* Chicago: University of Chicago Press. 1963 Phoenix paperback. A theory of hierarchical perceptual organization predating mine, and containing many ideas I have used (without knowing they were Hayek's).
Jones, R.W. 1973. *Principles of Biological Regulation.* New York: Academic Press. A very thorough introduction to biological control systems of all kinds, with many examples and full treatment of reference signals, perceptual functions, and output processes (not in my terminology, but just as good).
McFarland, D.J. 1971. *Feedback Mechanisms in Animal Behaviour.* London and New York: Academic Press. An often-cited book containing a wealth of basic examples, but little in the way of useful models. Nevertheless, a basic source.
Stevens, C.F. 1966. *Neurophysiology: A Primer.* New York: Wiley. A close look at neural functioning which is still essentially up to date. A strong antidote for the belief that neural processes are digital in nature.

[1979]

Degrees of Freedom in Social Interactions

Introduction

Freedom is well-known to be a relative term; one is not free to speak lies to a Grand Jury; one is not free to worship a god that demands human sacrifices; one is not free to publish facsimiles of U.S. currency; and one is not free to choose his domicile from those owned and occupied by others. We are all familiar with the fact that there are different degrees of freedom, none of them absolute.

The term "degrees of freedom" has another meaning, which turns out also to be relevant in this discussion. A physical system is said to have n degrees of freedom if n variables have to be given specific values in order to describe completely the state of the system at a given moment. An object in space has at least six degrees of freedom: three which relate to its location in a three-dimensional coordinate system, and three which relate to its orientation relative to the directions of the coordinate axes. If a system has n degrees of freedom, and all but one of them are specified by being given numerical values, there is only one way left in which the system can change. If the location of an object on a flat surface is specified in terms of an X-Y coordinate system, the only remaining way the object can move is up and down. If the X-position is specified and the height above the plane is specified, the only remaining degree of freedom is in the Y-direction.

Copyright 1979 by Gordon and Breach Science Publishers, Inc. Reprinted with permission of the publisher from Klaus Krippendorff, ed., *Communication and Control in Society*, 1979, 267-278.

In a model of a behaving organism, at least as I have approached the problem of modeling, these two senses of "degrees of freedom" turn out to be nearly the same in meaning. We are not absolutely free to indulge in certain behaviors because, I propose, neither we nor the environments with which we effectively deal possess an infinite number of degrees of freedom in the mathematical-physical sense. The limitation on freedom of behavior due to limits on degrees of freedom of organization can be seen in an individual, and in a society made up of individuals who interact with one another. As will be shown, mere masses of people do not create correspondingly large numbers of degrees of freedom in any sense; in fact, quite the opposite can occur.

Control Hierarchies

To begin at the beginning, let us consider a model of behavioral organization derived from Norbert Wiener's original insights [8], the later contributions of W. Ross Ashby [1], a seven-year collaboration with R.K. Clark [7], and the general literature of control theory and early engineering psychology. From these sources, and no doubt many others equally deserving of mention but too diffuse to recover, a picture has emerged of organisms as hierarchies of negative feedback control systems. This model is compatible with the thrust of Donald T. Campbell's long-term development of blind variation and selective retention, which has lately come close to a general control model of evolutionary and learning processes [3], but we will not here consider the origins of the organizations involved in fully-developed adult behavior. The main task here is to develop some general concepts which emerge from the control model, and see what implications they may have with regard to individual organization, and then social organizations.

A control system is defined in this model as any system capable of stabilizing against disturbance an inner representation of an external state of affairs. For all but practicing control engineers, that definition will require some elaboration.

First of all, what does it mean to say that something is stabilized against disturbance? This can only mean that this something, which we will call a *controlled quantity*, is affected both by independent influences and by the actions of the system itself, and that the system's actions systematically oppose the effects of disturbances on the controlled quantity. It was discovered, early in the game of control theory, that if a system is to stabilize some quantity it must sense that quantity, and furthermore it must have an internal standard against which to compare the outcome of that sensing process—a reference with respect to which the sensed quantity can be judged as too little, just right, or too much. The action of the system is based on that judgement, not on the sensed quantity itself nor on the reference itself nor on the disturbance. Departures of the controlled quantity from the reference level are what lead to the actions that limit those departures to a small or even negligible size.

Of especial interest to a control theoretician's thinking about behavior are some new interpretations of Piaget's work with children; von Glasersfeld [4] in particular seems to have found in Piaget an epistemological position that recognizes what I term "internal representation of an external state of affairs" as a central factor in behavioral organization, and furthermore recognizes that these internal representations, which we often term perceptions, are in all likelihood *arbitrary* transformations of the external state of affairs—"reality." In a control model of behavior this epistemological position thrusts itself upon one; there seems to be no other choice. A control system controls what is senses, and what it senses is the result of applying a continuous transformation process to the elementary sensory inputs to the nervous system.

One simple way to put this is to say that a control system controls some particular *aspect* of its environment. How many aspects might there be of a given environment? As many aspects as there are different ways of combining elementary sensory stimuli, and that is a very large number.

It is not, however, an infinite number. In fact, for a single organism, it is probably not even in the realm of what we call today large numbers. The number of degrees of freedom in the perceptual world is limited, of course, by the number of degrees of freedom in the physical universe outside, but it is much more severely limited than that: it is limited by the number of different aspects of the environment that a given organism is prepared to sense at a given time.

That is getting ahead of the story. At the moment, the important notion is that each control system, or subsystem within the organism, *constructs* a quantity by means of sensors and perceptual computations, and acts to stabilize that quantity, a perception. It acts to maintain that quantity matching an internal reference. If the reference remains constant, actions will simply oppose the effects of disturbances in the controlled quantity. If the reference varies, actions will automatically undergo further changes which cause the perception to *track* the changing reference. Control, therefore, may involve maintaining a static or a dynamic condition, protecting it against arbitrary influences from outside.

Already we have a picture of behavior that is considerably different from the traditional one. Instead of being caused to act by external stimuli, the organism under this control model acts purposively, the purpose being to maintain some perception, some aspect of the sensed environment, at a reference level specified inside the organism. There is much more to say about the model itself [6], but for the present we will simply adopt it, and elaborate on it to make it serve the purpose at hand.

A hierarchical control model can be constructed. It is possible to define a set of first-order control systems, corresponding primarily to spinal reflexes, which control very elementary aspects of the environment and musculature. There are probably hundreds of such systems, all at the same level. Each such system requires a signal from further inside the organism which will set its reference level—tell it how much of its perception to produce and maintain. The controlled perception, together with many other perceptions, controlled and uncontrolled, is available to systems further inside the organism. Thus the systems further inside have second-order sensory information about the external state of affairs, and can act to change those second-order perceptions by adjusting the reference levels of the first-order systems which actually create muscle forces in the process of their own control actions.

In this way a hierarchy is built up, a system at a given level acting by adjusting the reference levels for lower-level systems, and acting for the purpose of stabilizing its own representation of the external state of affairs.

Now we can begin to see how the degrees of freedom of action might be constrained to a relatively small number at any given time. By "action," we must mean in general not motor behaviors, but selection or specification of reference levels for control systems. If there are n independent control systems operating at a given level in this hierarchy, then there are just n reference levels that can be independently specified—at most n, assuming no limitations introduced at lower levels.

It is not necessary to think of one lower-level system having its reference specified by just one higher-level system. Much more generally, and probably more realistically, we can think of the set of all reference-inputs to lower-level systems as defining the degrees of freedom for action, and we can let any higher-order system act by contributing to the reference-settings of many lower-order systems at once. In this way we could accommodate n higher-order systems acting at once on a large subset of m lower-order systems, any one lower-order system receiving a net reference-setting that is the sum of influences from many—perhaps even all—of the systems in the level above. As long as the n higher-level systems distribute their contributions to the lower-level reference-settings properly, no conflicts will be created (no paradoxes), and each of the n higher-order systems can act independently to control its own perception.

The crucial word is "properly." The proper distribution of effects can be found only by solving a set of simultaneous equations, and if there are n systems cooperatively making use of n control subsystems, there is one and only one possible solution to the system of equations.

When control systems are involved in situations like this the problem is in some regards not so difficult to solve as it would be if all the systems operated blindly. In fact, the requirements on the distribution of actions by higher on lower systems are much relaxed; some considerable degree of interaction is permitted, the actions of one system disturbing control processes in other systems of the same level. The ability of control systems to maintain control in the presence of disturbances also permits them to work when coupled together—not too tightly—at their outputs.

But the restrictions are still severe enough. When control is involved, the requirement shifts to the *input* side; now it is necessary that each higher-level system sense a different aspect of the lower-environment (consisting of lower-order perceptual representations). By different aspect we mean, of course, an independent aspect; one which could freely be varied no matter what states were being maintained for other aspects of the same environment. So now we are talking about degrees of freedom of perception instead of action, but the basic requirement has not been fundamentally changed: the set of simultaneous equations must still, somehow, be solved, if the n higher-order systems are all to be capable of independent action.

Perceptions are neither entirely consistent nor entirely free of random variations; furthermore, the human brain does not become organized through systematic solution of equations, but through variations (blind or otherwise) and retention of resulting organizations that work to the organism's advantage. It is thus highly unlikely that any hierarchically-organized organism would prove able to employ all n degrees of freedom inherent in an assemblage of n control systems at the same level, even assuming that levels are neatly separated in nature. In order for successful control organizations to emerge, it is most likely the case that the number of available degrees of freedom must be far larger than the number ever simultaneously employed.

The basic reason why these "equations" have to be "solved" is that control organizations cannot work under conditions of direct conflict; the better the quality of control, the worse the consequences of conflict. Failure to maintain independence of the aspects of the environment under control shows up as an attempt by two systems to control the same quantity relative to two different references, which is impossible. Failure to solve the equations thus means loss of control. Loss of control means lowering of survival potential, so we can expect that evolutionary processes will have selected against significant amounts of internal conflict.

Now we have a general picture of a hierarchically-organized organism based on control principles, and we can see that at each level in the hierarchy there is a problem of degrees of freedom any time that many higher-order systems act simultaneously on and through the same set of lower-order systems. The problem is simply that of avoiding internal conflict and losing control altogether. We can now put this aspect of "degrees of freedom" together with the other aspect, the idea that freedom is never absolute, in the less technical usage of the term. Then we will be ready to look at social systems.

Relative Freedom and Relative Purpose

When it is said that we are not free to shout "Fire!" in a crowded theater, there is an unspoken assumption in the background. We are not free to do this *if* we are aware of and reject the alternatives: causing a fatal panic or going to jail. Of course if we are willing to accept the consequences, or are unaware of any, we are perfectly free to shout what we like where we like. It is not any law of nature that limits our freedom in such circumstances; the only limit is imposed by the organization of the person in question.

More specifically, the limitation of freedom is imposed by the fact that doing certain acts that satisfy one set of goals or purposes (or in the more noncommittal language of control theory, reference levels) can cause other controlled perceptions to depart from their reference levels. Conflict can be created, depending on one's structure of perceptions and the reference levels that go with them. If a person wishes strongly to avoid going to jail and also wishes strongly to shout "Fire!" in a crowded theater, he has a problem. It is not a problem of sequencing; it is a direct conflict, in that satisfying both reference levels (matching present-time perceptions to them) is essentially impossible. The impossibility is what limits freedom.

Suppose that someone possesses n independent control systems at some level in his hierarchy. This means that in principle, this person could set reference levels for each of the n perceptions in any way he liked. But now suppose that he has, for whatever reasons, already specified $n - 1$ reference levels, thus implying that $n - 1$ aspects of the lower-order environment (seen from this level) are being controlled in specific states.

How much freedom is left for setting the remaining reference level?

From one point of view the remaining goal could be chosen with complete independence; the system setting that goal is not constrained. But there is only one value that can be specified for the remaining reference level that will not create direct conflict. Direct conflict will destroy the ability to control, or severely impair the control abilities of all the systems involved. This will not directly affect the system that erroneously assumed freedom to set the remaining reference level, but there will be a definite indirect effect, if we are talking about a hierarchical control organization. The higher-level systems set lower-level references as their means of control; if the lower systems come into conflict, the result will be an impairment of higher-order control processes. Thus there is a purely *mathematical* limitation on the freedom of the whole system to set simultaneous reference levels at a given level in the hierarchy. Attempting to violate that limitation destroys control.

Reference levels constitute intended perceptions: purposes. In a hierarchical system, any level of purpose is at the same time a means for achieving higher-level purposes. The runner adjusts the speed with which he intends to run as a means for controlling the place in which he will finish the race; if it is impossible to come in first, a different choice of running-speeds may make second place possible, and a poor choice will assure not finishing the race at all. Purposes generally have this quantitative aspect in addition to their qualitative natures. Clearly, one is not free to select purposes at a given level if those purposes are also means for achieving higher-order purposes.

In fact, when there are higher-order purposes, external circumstances can be as important a determinant of selection of purposes as any internal determinism at higher levels. If one intends to enter a house, the selection of lower-order controlled perceptions that is made will be strongly influenced by the state of the front door: open or closed, locked or unlocked. A successful control system adjusts its actions to oppose external circumstances that would tend to disturb the controlled quantity—which means that it must allow external circumstances, to a great degree, to control its actions, its selection of lower-level goals.

When freedom and purpose are examined in the light of this hierarchical control model, we can see that no simple slogan can unravel the complexities that come up. "Free will" is a phrase that would be used only by someone who has not really thought about the whole problem. Freedom is relative; sometimes it is impossible. Our chief freedom, it seems, may be the freedom to seek the state in which we suffer the least internal conflict, and thus remain capable of acting on the environment in the way that lets us continue functioning according to our own inner requirements, whatever the basic requirements may be. I would not rule out of this set of basic requirements, by the way, concepts such as orderliness, beauty, elegance, or progress. We have not yet read the entire message in the genes, nor are we in a position to put limits on what it might imply.

Freedom in Social Interactions

We have been discussing so far only an individual person. The principles developed, however, should apply at least in part to a social organization, because we have dealt with the individual as a collection of control systems that must work in harmony in order for the whole to function properly. This certainly suggests a parallel treatment of a collection of individuals, each of whom acts as a control system in any given circumstance, and all of whom must learn to live in harmony to avoid the dangers of social conflict.

Before developing these parallels, I want to warn against one tempting extension of the control model. In a human being, it is possible to identify each part of a control system with some part of the person—his sensory equipment, cerebral processes, motor organization. Every function that has to be performed to make up a control system has a place to live, a place where it can be embodied in tissue. When the same organization is extended to a social system, such as an army, one can see many counterparts, but on close inspection these counterparts to the components of a control system are no more than metaphors. An army does not have a perceptual organization, a way of making comparisons, a motor output system. Only the people in the army have such things. An attempt to transfer a control model of an individual directly to a social organization violates the kind of model-making spirit that demands a relationship to the physical world. The kind of model that is needed to represent the interactions among the individuals who make up an army may prove to include feedback phenomena and many other features of a control model —but an army will never be a control system in the same sense that an individual is.

What can we say about social interactions without depending on loose metaphors? And also, I should ask, without having done the many lines of research implied by this approach? Whatever is said will obviously have to be general and tentative, but there are some useful implications to consider.

The existence of conflict between individuals is of theoretical interest if only because of epistemological implications. Although each person might be acting primarily to control aspects of a constructed perceptual reality, the fact that interpersonal conflict can exist and persist indicates that there are regular objective consequences of control behavior. Even though the world of controlled quantities is primarily subjective, it seems that control actions entail producing regular effects in the outside world, regular enough that the control behaviors of two persons may prove mutually exclusive. For one person to maintain control of his perceptions is for another to lose control of his. That is the essence of interpersonal conflict.

We live in a society in which competition is praised as a spur to greater efforts and a higher quality of production. The free enterprise system in principle permits each person to look after his own interests, advancing himself relative to others by increasing the value of his work to others. Again in principle this process should lead to the state in which each person has found a niche which maximizes the benefit to himself while at the same time maximizing his contribution to the well-being of all other persons. Our educational procedures, through competitive sports, grading on a competitive and comparative basis, and at the higher levels acceptance into schools on the basis of relative accomplishments, emphasize and nurture this competitive concept of social evolution.

[Handwritten at top: "This is extremely profound & applies a model for solution of A. Kuhn's theories"]

When sufficient degrees of freedom exist, this design for a social system would seem quite feasible. It is, in fact, a heuristic that leads to the minimization rather than the emphasis of interpersonal conflict. If the system really worked as it is imagined by its proponents to work, *direct* conflict would result in a winner and a loser, and the loser would turn to some other endeavor. Eventually each person would find a position in the society in which his own structure of reference levels could be satisfied without excessive effort; since the principal cause of excessive effort is direct conflict with other controlling organisms, the result would be a minimization of direct conflict, and in effect a solution to the system of equations describing each organism's requirements and the set of all available means for meeting those requirements. Thus the result of optimal functioning of a society ostensibly based on the principle of free competition should be the reduction of direct competition to the minimum that is actually achievable [Kuhn, 5: 216 ff].

There are at least two important factors which prevent this system from operating to reach a low level of conflict. The first is the fact that a person in a society can interact with the society only as he understands it, not as it actually is. It is not generally understood that the final outcome of free competition ought to be a minimization of competition. Instead, the way we train ourselves to live with competition has resulted in a glorification of conflict itself. Rather than trying to maximize the useful product of our labors, through reduction of conflicts which produce mutually-opposing efforts, we have come to make the opposition of one person by another into a source of reward and prestige. The achievement of such rewards and such prestige becomes a goal accepted as important by large numbers of people, even though few of them have any realistic hope of joining the small circle of winners. Entering into conflicts is said to strengthen us, although in most such cases, such as the example of professional football, it would be hard to say how the resulting kind of strength would be of use to us except in the conflict situation itself. In a society that accepts conflicts as something to be sought, little value is placed on the cleverness with which we find social solutions that avoid direct conflict. Each person's understanding of the nature of his society and the goals implicit in that society organizes all his lower-level purposes. If those purposes do not include the minimization of direct conflict, it becomes rather unlikely that we will find a peaceable solution to our social problems.

The fact that our society apparently seeks conflict situations is not, in the long run, the most important impediment to getting rid of debilitating and dangerous conflicts. Whatever theoretical notions a person may have about his society, he is not going to continue entering into conflicts deliberately if the result is consistently against his interests. There will be at least a strong bias toward avoiding conflicts in which one runs a high risk of emerging the loser. The older one gets, the more evident it becomes that the thrill of victory is ephemeral; while the agony and consequences of defeat are cumulative. One cannot afford to go on forever using his efforts merely to oppose the efforts of others; at least a major part of one's effort ought to be directed toward assuring one's own survival. One cannot take care of himself or his family if the major part of his effort is cancelled out by the efforts of others trying to counteract his disturbances. Sooner or later, one looks for the path of least, not most, resistance.

Then the question becomes: does such a path exist? The answer depends on degrees of freedom.

The number of degrees of freedom in the physical environment is, according to what physics would say, inexhaustible. But it is not a physical model of the environment with which we normally interact and within which we choose our purposes. It is a perceptual model within which we find our goals. The worlds we attempt to control relative to our goals, and the goals themselves, are made up of automobiles and hamburgers, jobs and vacations, bowling and cross-country skiing, passing algebra and plying ladies with gifts in the effort to overcome resistance. It is almost entirely a manufactured world, a world divided into familiar perceptual categories and familiar examples of each category. It is a rather small world, the smaller to the extent that we come to share more and more classes and examples of perceptions, rather than creating our own categories and examples.

This is the other side of civilization. On the one hand, by banding together and pooling our efforts, we can achieve for all of us what none could achieve for himself. On the other hand, by banding together and creating a shared reality, we reduce the size of the universe in which we live, narrowing the choices of goals and the actions recognized as means toward goal achievement. The more of us there are, and the more closely-knit the society we perceive and accept, the fewer become the unused degrees of freedom and the higher becomes the likelihood of direct conflict. The final result can only be a society in which for each person there is one and only one conflict-free set of goals possible, at every level of his organization. All freedom of choice vanishes.

Within an individual, as already mentioned, the hierarchy of systems comes into being through variations and retention of profitable results, not through systematic and efficient design processes. The same holds true for a society. And just as inside one person, in a social system it is not feasible to match the number of personal goals to the available degrees of freedom. The world in which we live—the effective world, the one we perceive—must have far more degrees of freedom than we have goals, if we are to hope to approach something like a minimum level of conflict. If all degrees of freedom but one were exhausted, it is not likely that the remaining one would ever be discovered. Our goal structures must be such that there are many actions that would serve to satisfy any given goal; the richer the store of alternatives, the more likely we are to be able to minimize conflict and maintain control.

The implications are obvious. The more standardized a society becomes, the fewer become the individual goals and the means for achieving them. The more people there are, the fewer degrees of freedom remain. Long before actual exhaustion of degrees of freedom occurs, the level of conflict within a growing and increasingly standardized society must begin a rapid ascent. Failure of an individual to find a unique set of goals and an unopposed means to achieve them forces that individual to compete with others for means and to select goals which can only be met if someone else fails to maintain control. Finding the unique set becomes difficult long before the last degree of freedom is used up. Overpopulation and overstandardization begin to have their effects long before they are recognized as such. The symptom is not any dramatic confrontation among individuals. It is simply an increasing amount of difficulty experienced by everyone in going about his affairs. If too many people decide to take up macramé, each person will find his local store low on the supplies he needs. It is as simple as that. We begin stumbling over each others' feet long before we realize that there are too many of us in the same place trying to do the same things.

Conclusions

Control theory throws a new light on the subject of conflict, whether it be interpersonal or intrapersonal. When two independent control organizations come into conflict, the result is not simply a vector addition of the efforts. It is an abrupt increase in the efforts, most of the increase of one system's efforts serving only to cancel the increase in the other system's efforts, and producing no useful result for either system (except, perhaps, an increase in the volume of muscle tissue). While it is true that in the process of resolving conflicts the participants may develop new abilities, the glorification of conflict will not tend to develop abilities that are of general usefulness when there is no conflict. It will result only in hypertrophy of some function out of all proportion to the others. The glorification of conflict results in the Muscle Beach syndrome. The abilities developed through prolonged conflict generally go to waste unless another similar conflict can be found. Thus conflict as a goal will not stand up to analysis in terms of a hierarchy of goals.

Once the peculiar disadvantages of conflict between control systems are grasped, it is seen that conflict is the key to understanding many social problems. This conclusion is, of course, quite in line with common sense and experience, and by itself is nothing earth-shaking. But control theory shows us in great detail just *why* conflict has bad effects, and it leads us to see a relationship among conflict, overstandardization, and overpopulation, a relationship that has long been intuitively obvious but which now assumes the proportions of a natural law. Whatever our beliefs about the benefits of conflict in driving people to greater achievements, control theory makes it clear that conflict itself cannot be good for us, any more than breaking a leg is good for us just because it exercises our self-repair machinery. It is true that in the course of trying to resolve conflicts, trying to find a solution to the equations of life, we often come across new skills and knowledge that remain of permanent value when conflict is removed. But we should not forget that there are many ways to accomplish any given purpose, in a rich enough and sparsely-enough populated environment, and our goal is to be able to accomplish our purposes, not deliberately to seek impediments.

Through an understanding of social systems in terms of individual control systems acting independently of each other, I think we can arrive eventually at some clear statements about what is going wrong in the world we share, and perhaps even begin to see a way out of some major problems. A few cherished beliefs may become casualties, but if we can come to understand the real reasons behind the increasing tension and violence of our world, and to see that the main problems arise from attempts to violate immutable laws of nature and logic, we should not find incorrect beliefs too hard to abandon.

References

[1] Ashby, W.R., *Design for a Brain*, New York: Wiley, 1952.
[2] Campbell, D.T., "Social attitudes and other acquired behavioral dispositions," in Koch, S. (Ed.), *Psychology: A Study of a Science*, Volume 6, *Investigations of Man as Socius*: 74-171, New York: McGraw-Hill, 1963.
[3] Campbell, D.T., "Downward causation in hierarchical selective-retention and/or feedback systems," Paper presented at the Symposium on Cybernetics in Psychology, American Psychological Association Convention, Washington, D.C., September, 1976.
[4] Glasersfeld, E. von, "Piaget and the radical constructivist epistemology," in Smock, C.D. and Glasersfeld, E. von (Eds.), *Epistemology and Education*: 1-26, Athens, Ga.: Mathemagenic Activities Program—Follow Through; Univ. of Georgia, 1974.
[5] Kuhn, A., *The Logic of Social Systems*, San Francisco: Jossey-Bass, 1974.
[6] Powers, W.T., *Behavior: The Control of Perception*, Chicago: Aldine, 1973.
[7] Powers, W.T., Clark, R.K., and McFarland, R.L., "A general feedback theory of human behavior," Part I, *Perceptual and Motor Skills*, Monograph Supplement, 11, No. 1: 71-88, 1960; Part II, *Perceptual and Motor Skills*, Monograph Supplement, 11, No. 3: 309-323, 1960.
[8] Wiener, N., *Cybernetics*, New York: Wiley, 1948.

[1986]
On Purpose

The concept of purpose has been in bad repute among life scientists since before they adopted that name. Control theory, on the other hand, shows that the principal property of organismic behavior is its purposiveness. There is clearly a problem of acceptance here, especially because anyone who speaks of purpose in polite scientific company is likely to detect a common reaction—oh, you're one of *those*. The difficulty is that the word purpose evokes images of mysticism and religious persecution, throwing the whole discussion into the wrong category. It's hard to persuade a scientist to take another look at the phenomenon when he or she is convinced that there isn't any phenomenon.

What is the phenomenon? It can be described very simply. We observe an organism in its natural habitat over some period of time. We see that it carries out typical behaviors again and again, maintaining itself in certain relationships with its environment and causing repeatable effects on its environment. It might seem at first that these regularities could be studied in the same way we learn about ocean currents, orbits, and crop yields: by finding the antecedent conditions that govern the observed behaviors. Actually this approach works very poorly; we are hard-pressed to find even statistical regularities. In trying to understand why behavior is so variable, we look closely at the details, and find a puzzle. While the general outcomes of behavior often repeat well enough for us to study them, the actions that bring about these outcomes

Copyright 1986 by William T. Powers. Reprinted from *Continuing the Conversation* (7), Winter 1986, 1-3.

vary almost at random. There's a kink in the causal chain.

One way to eliminate the kink is to do experiments under controlled conditions where a given outcome can be brought about by only one action. In that case the chain straightens out and the same action always leads to the same result. If the only way the rat can get a food pellet is to press a bar, it presses the bar. Of course it might do so with any of four legs, while facing toward or away from the bar, or it might nose it down or sit down on it, but it always presses the bar at least hard enough to make the response-recording contact close.

While those controlled conditions make life easier for the experimenter, the experimental animal doesn't need that much coddling. Organisms produce specific outcomes, not specific actions: their actions adjust according to the momentary requirements of the environment, so that when all the influences on the outcome are added up (including the influences created by the organism), the same result appears. That is what makes behavior seem purposive. Organisms don't just go through the motions like automata; they vary their actions in whatever way is needed to achieve the results we recognize as behavior. It seems that they produce those outcomes on purpose: that they intend that specific outcomes will occur, and vary their actions in any way needed to bring about those outcomes or maintain them against disturbances.

When you think of a behavioral outcome as a physicist would, you see immediately that actions MUST vary if that outcome is to repeat. That is because other forces and constraints are always acting on the same outcome. If the independent influences change but the outcome doesn't, physics demands and reason deduces that the action must have changed, too, precisely and quantitatively the correct way. Observation confirms this expectation in essentially every instance of behavior.

Neither physics nor reason is influenced by mere beliefs: if actions systematically oppose disturbances, that is all there is to it, they do. There is then nothing to keep an engineer using physics and reason from wondering how a system has to be organized to behave that way, discovering how, and building some examples to learn more about the principles of such organizations. The engineers who did that invented servomechanisms, and the principles they developed are called control theory, the foundation of Cybernetics.

Life scientists, however, didn't take that approach. Instead of finding out how purposive behavior works, they decided that it doesn't exist. Most of them simply ignored the kink in the causal chain: they experimented and reasoned and explained just as if there were no kink, as if regular outcomes are produced by regular actions. Among the few who decided not to ignore the kink, many explanations were offered to make it seem that the kink didn't matter. Some even decided that behavior really isn't regular, that we just classify random outcomes by similarity, imposing our desire for order on a basically random phenomenon. This preserves regularity in a different way, by asserting that irregular actions can lead only to irregular behavior, a blithe denial of observation.

Most of these scientists studied behavior as if regular outcomes were caused by regular antecedent conditions. They may have been working with real organisms, but what they saw was the organism they believed in. Belief feeds on confirmation and ignores counterexamples. What better aid to maintaining beliefs is there, than statistics? By observing relationships among antecedents and consequents statistically, one can ignore the precise and systematic variations of action that make the outcomes regular, in effect looking at behavior under the influence of the *average* disturbance and the *average* purpose. The essential data that prove behavior to be purposive are thus discarded as statistical noise.

Once it was agreed that purpose is the figment of a primitive imagination, it became the duty of every life scientist to explain behavior without taking purpose into account. You may ask, "How could they do that, being scientists, if behavior really is purposive?" The answer is, easily. All you have to do is keep in mind that behavior is caused by what happens to an organism, and not by any purposes inside the organism. You then vary your interpretations and observations until they make that fact true. Here is a brooding bird removed from its nest: it struggles mightily in the direction of the nest (no matter how far or in what direction from the nest you have taken it). What makes it do that? Why, the sight of the nest, what else? The visual image of the nest acts on the retina and nervous system, causing the muscles to produce forces in the direction of the nest. If a physicist or an engineer listens to this explanation with jaw agape, the behaviorist listens with satisfaction: it keeps faith with the basic premise of external causation, which is more important than asking a lot of finicky questions like "how?" Since we know behavior is externally caused, we don't have to take every fussy little fact into account. How often do we have to demonstrate an established principle?

I would settle for once.

Naturally, the custom of bending reason to accommodate a preselected premise has not encouraged clear thinking. Consider the subject of reward or reinforcement, and its relationship to the behavior that produces it. In an operant-conditioning experiment, a "contingency" is established through a "schedule of reinforcements," usually embodied in an apparatus that converts behavior into delivery of food pellets or some such valuable objects. Behaviorists have used such experiments to show how the reinforcements maintain the behavior that produces them, and how the details of the schedule influence the rate and form of behavior. They claim that they have shown by direct experiment how external circumstances control behavior: just the facts, no theory.

When you hear the word "schedule," don't you think of something like a timetable of events, like train departures or movie showings? It wouldn't be hard to see how a schedule of train departures would affect the times at which you might show up at the train station, and in fact you might admit that the schedule essentially determines when you will go, and to what station (if your purpose is to take the train somewhere). From hearing behaviorists talk, you might get the idea that a schedule of reinforcements works the same way: some routine for administering reinforcements is laid out in advance, and from knowing the schedule, one predicts the behavior.

That isn't how it works. A typical simple schedule could be described this way: for every tenth press of the lever, one pellet of food will be delivered. Does that tell us anything about when or how often food pellets will be delivered? If the organism never presses the lever there will never be any food delivered. The number of pellets delivered will be one tenth of the number of presses of the lever. The organism could press the lever in any pattern and at any rate whatsoever, and the reinforcements would dutifully appear at a corresponding rate and in a corresponding pattern. The ratio of reinforcements to lever presses is determined by the apparatus, but nothing else is determined.

In fact, in order to predict what the actual schedule of reinforcement will be, one would have to know what the actual pattern of behavior will be: the reinforcement depends entirely on the behavior, according to the settings in the intervening apparatus. This dependence is directly observable.

How do we get this simple relationship turned around to make reinforcement the cause and behavior the effect? Very simple: we turn to an insufficiently neglected mode of argument called assertion. We KNOW that the causes are external, however appearances, reason, and physics might delude us.

Lest anyone feel too superior to the behaviorists, consider this question. Do you think that you can make a child behave better by giving the child rewards for good behavior? Most people, I think, would say, "Of course." Actually, experience in many cases would justify the answer. But given that, how many of you would than say that the child behaved better *because of* being rewarded? Aha. Most of you.

But stop and think. What if the reward you used was a yummy tablespoon of vinegar? Oh, well, that's not a reward, you say. But why isn't it? What makes this stuff a reward when it's given, that stuff the opposite, or neutral (a yummy tablespoon of water)? At this precise point you switch from thinking of YOURSELF as the cause of behavior to thinking from the point of view of the child. A child, you point out patiently, wouldn't go to any trouble to get a tablespoon of water, and would probably go to a lot of trouble NOT to get a tablespoon of vinegar. I play dumb, and ask, well then why would the child go to any trouble to get a tablespoon of chocolate syrup? You explain, perhaps not so patiently, "The child LIKES chocolate syrup, dummy!"

Oh, I see. The child is changing behaviors so as to get a tablespoon of chocolate syrup, is that it? Say yes, this is a Socratic Dialogue. And the child knows that if it behaves in a certain way you are likely to whip out the Hershey's? Say yes again, unsuspectingly. NOW I HAVE YOU. The child is acting in a certain way *in order to make chocolate syrup appear*, taking advantage of the fact that you are obeying a reliable rule for delivering the reward. The purpose of the child's behavior is to control the delivery of chocolate syrup, right?

Oh, no, you don't! Behavior isn't purposive! It's caused from outside! You have it all backward! You tricked me!

This Socratic Dialogue has gotten out of hand, as real ones do, but you get the point. You don't cause behavior by giving rewards. You just put yourself in the position of being used by someone who knows how to get you to give what that person wants. This is a perfectly good way to get people to use certain actions to get what they want, and maybe you can even teach them something in this way, but if the person doesn't want what you have to offer, you might as well give up. If you do give up, the person is likely to find another way to get the same thing: the actions aren't important. The result is the point: the result the person intends to get.

The human race has been using words like intention and purpose for a long time without any idea of how such things could exist (it did the same thing for equally long with words like "digestion," I should add). As we tend to do with all unexplained phenomena, people have tried to make sense of purpose and intention, whether or not they had any means of doing so. The result has been a great many flights of fancy, basically no more meritorious than fancying that purpose and intention don't exist. The arguments on both sides of this issue have necessarily been based on ignorance, because the means of understanding, control theory, wasn't worked out until the mid-1930s. No argument about purpose prior to that time could possibly have made any sense, however close it may seem to have come to the proper explanation.

The consequence of this long period of argument ex vacuo is that some positions have been very firmly established, for no good reason. Anyone who steps in now and offers to show how purpose really works is likely to be rejected by both sides: the proponents say it can't be that simple, and the debunkers think you are arguing on the side of the proponents. If you're not a Big-Endian, you must be a Little-Endian.

When you learn control theory you learn what a purpose is

and how it works. You strip away irrelevant issues such as verbalization, consciousness, complexity, and position on the evolutionary scale. Life is purposive at every level of organization, right down to the little enzymes hopping along the backbones of DNA molecules, repairing them. Purpose is an inherent aspect of the organization known as a control system. It works just as people have always thought it works—without the metaphysical baggage. We might as well start using the term freely, because it's here to stay.

[1987]

Control Theory and Cybernetics

In recent CCs, there have been some self-promoting complaints about how unaesthetic we control theorists are. From the receiving end, this is something like getting an obscene phone call: it's hard to think of it as a conversation. Well, I won't put up much of a defense. There *are* some dull spots to get through on the way to understanding control theory, and a control theorist would be the last person to say anyone *has* to like control theory, or understand it. On the other hand, if you don't understand control theory, isn't it a little unwise to write thousands of words about what you imagine it to be? I would think that the potential for embarrassment would be reason for caution.

The control theorist isn't trying to reduce human beings to machines, or trying to draw clever analogies between human activities and those of Rube Goldberg (or Bucky Fuller or Department of Defense) artifacts. Instead, he or she is trying to make a start on understanding human nature and the nature of organisms in general in some useful way. This has never been done before. Perhaps some cyberneticists, despite their assessment of the state of the world, don't like to hear statements like that. I can assure you that conventional behavioral scientists don't like to hear them, either. Control theorists have had just as hard a time with conventional behavioral scientists as they seem to be having with certain cyberneticists, and for similar reasons: the opposition is arguing against something they haven't taken the trouble to understand.

Copyright 1987 by William T. Powers. Reprinted from *Continuing the Conversation* (11), Winter 1987, 13-14.

Behavioral scientists like to discount the successes of physics and engineering by saying that the hard scientists have it easy: they work with reproducible phenomena and simple material objects, whereas students of living systems face immense complexity and variability that call for a different approach. Basically, that is hogwash. If organisms are so complicated (and certainly they are), why is it that the analyses of their behavior offered by behavioral scientists are so utterly simple? Most explanations of behavior can be reduced to the statement "Behavior B is caused by stimulus, situation, cognition, or property A." Now compare that kind of analysis with the kind a physics student struggles to understand while learning to predict the behavior of a simple piece of matter, a spinning gyroscope. Is it easier to get an "A" in Physics 301 or in Psychology 301?

The reason that the behavioral sciences have had so little success is twofold: first, the aim is wrong, and second, the model is wrong. The avowed aim of behavioral science, in many quarters, is the "prediction and control of behavior." This goal makes sense only in terms of a model that describes behavior as an effect of external causes (and not of goals). If the scientist can study external causes and the behaviors they generate, it follows from this model that by observing or predicting new circumstances, the scientist can predict new behavior. And most important, by manipulating those circumstances, the scientist can control behavior.

Control theory shows that the cause-effect model is wrong, and therefore that the goal of predicting and controlling behavior is trivial, futile, or self-defeating. For a lot of detailed reasons that I won't go into here, because they are somewhat dull and space is limited, the control theorist understands behavior as the process by which organisms control the worlds they experience. The standards around which this control process is organized are inside the organisms, not outside them. Control systems in organisms take on specific forms through interactions with the world outside, but they also reflect inborn organization that can't be traced to any event or cause in the lifetime of a single organism. There are basic goals, intentions, standards—we call them "intrinsic reference signals" to get away from old meanings and to distinguish them from learned goals—that define for us what it is to be human: that tell us that we will find experiences pleasant or painful, aesthetic or ugly, orderly or chaotic. At the lowest levels, intrinsic reference signals determine what pH will be maintained in the bloodstream, what temperature in the brainstem, what level of lactic acid concentration in the muscles. The lowest known level in the intrinsic hierarchy seems to be found in the repair enzymes that are made by and act upon DNA. At the highest levels, intrinsic reference signals perhaps set the very terms in which we make human value judgements: the dimensions along which we judge, rather than the specific judgements.

In service of these fundamental standards or reference signals, we acquire through experience a hierarchy of behavioral control systems. These control systems exist, in an adult, at many levels, and in multitudes at any given level. They all operate at the same time, sometimes consciously but most often not. A few of these levels deal with symbol manipulations, but there are levels both higher and lower than these "rational" (meaning, mostly, verbal) levels. The higher levels operate by adjusting the goals of the lower levels, the lowest level in the behavioral hierarchy being the spinal "reflexes," and the highest I can think of, for the moment, being concerned with system concepts like self, society, science, and art (to name a few). The control system model thus sketches in the necessary steps of translating thought into action and vice versa.

That is, very roughly, how the control system model of behavioral organization is put together. Behind this model there is something called control theory. Control theory does not consist of the statement that organisms are control systems —that statement proposes only that certain relationships will be seen in behavior; if they are seen, the behavior is indisputably that of a control system. Control theory is the method of analysis that lets us understand and predict the behavior of any system in this kind of closed-loop relationship with an environment: basically, it's a body of mathematical analysis. In that respect, it's like Von Foerster's attempts to represent behavior in terms of recursive functions and eigenvalues, or Varela's use of the Spencer Brown calculus with the addition of strokes that go around little squares. The difference between the latter two approaches and the approach using control theory is that control theory actually makes quantitative predictions of real experimental data—very accurate predictions. The other two approaches have yet to predict any specific observable measure of behavior, accurately or otherwise.

From the standpoint of the conventional behavioral scientist, the control theoretic picture amounts to a total repudiation of the conventional concept of what behavior is and how it works. I am puzzled to find that cyberneticists, particularly the ilk inhabiting the pages of CC, have not greeted the advent of control theory with cries of joy. Control theory supports many of the objections to conventional science that are apparent in these pages—and slips a scientific foundation under them. Unfortunately, there have been many interpretations of control theory based on half-understood rules of thumb, and many leaps to wrong understandings of what control systems are and how they work, published in the cybernetic literature as well as elsewhere. So the objections directed at control theorists are mainly misdirected: they impute to us things we don't believe, they make wrong deductions from control theory and then object to them, and, if you'll pardon my pique, they sometimes reject what is really a very beautiful and precise concept while substituting a lot of empty holier-than-thou blather for it.

It's not really fair to argue against control theorists by imputing to them the beliefs, aspirations, and philosophical stances of the very sciences they are trying to revolutionize. The control theorist does not believe that "scientific method" as now used with respect to organisms is worth much. The control theorist is, true enough, concerned with quantitative analysis, but is also vitally concerned with human capacities for perceiving the qualities of experience, from simple intensity to system concepts. Imagination, insight, creativity, and feeling are all part of human nature, and we control theorists try (with varying success) to integrate them into our models. Control theory—*real* control theory, not that "programmable functions of stochastic machines" junk—probably gives us the best medium for understanding constructivism, for making it real, illustrating its premises, and saving it from solipsism. Control theory is exactly what cybernetics needs. That's not so strange: control theory is exactly what cybernetics was founded on, however many cyberneticists have forgotten that (or never knew it).

Some wise men of the East advocate a life of passive perception: go with the flow. Some wise men of the West advocate a life of blind action: damn the torpedoes. I don't think that the solution to human problems has been carried very far by either group. I hope that cyberneticists (and everyone else) will be able to accept a new approach to human nature that is based on the hard demands of good science, and, even if it offers little that is spectacular right now, will understand that to build a real science that can solve social problems, we must begin at the beginning. The truths of control theory are truths that work with precision, all of the time, admitting no exceptions. If they are simple truths, so be it: so were those that Galileo found by rolling balls down a ramp and timing them with his pulse. If they are provisional and temporary truths —well, name me a truth that isn't both of those things.

[1988]

The Asymmetry of Control

The circular relationship between organisms and environments is well known: behavior affects the environment and the environment affects behavior. On superficial consideration it may seem that we have a choice: the organism controls its environment, or equally well the environment controls the organism. This is not true.

To see that there is asymmetry in this relationship we can boil the situation down to its simplest elements. In Fig. 1 are two triangles representing agencies. The points are the outputs. The side opposite each point is the input surface, which receives two input effects. One effect is constant: the inputs labelled r and d. The other effect is simply the output of the other triangle, labelled respectively p and a. The output a is some constant K times the sum of inputs r and p, and the output p is another constant E times the sum of inputs a and d.

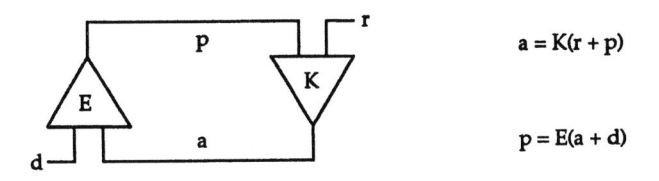

$$a = K(r + p)$$

$$p = E(a + d)$$

Figure 1.

Copyright 1988 by William T. Powers. Reprinted from *Control System Group Newsletter*, February 1988, 3.

We have a feedback situation. For this combination to be stable, the feedback must be negative, so we know immediately that K and E must have opposite signs. That is not, however, the asymmetry of which I speak, as either one can be negative. To see the asymmetry we must solve the system equations as a simultaneous pair, to get

$$a = \frac{KE}{1-KE}(d + r/E), \text{ and}$$

$$p = \frac{KE}{1-KE}(r + d/K).$$

If K and E are both very large numbers, one negative, then a = d and p = r. Each agency makes the other's output match the "loose" input, which is the reference signal. So each agency controls the output of the other, and there is symmetry. But if K is a very large number and E is around unity, the agency with K in it will make the other's output, p, match its own reference signal, r, but the other agency will not be able to maintain the same relationship.

The agency with K in it is the organism. Organisms are highly sensitive to inputs, but environments do not correspondingly amplify the inputs that affect them; normally there is a *loss* of effect: E is generally less than unity. The organism's reference signal r thus *does* affect the environment, while the environment's "reference signal"—the disturbance d—*does not* have a corresponding amount of effect on the organism. Organisms control environments, but not vice versa.

[1988]
An Outline of Control Theory

Nearly 100 years ago, William James pointed out that organisms differ from every other kind of natural system in one crucial regard: they produce consistent ends by variable means. He made this observation just at the dawn of so-called scientific psychology: his words were quickly forgotten. In their eagerness to make the study of behavior into a science, the American psychologists who became the intellectual leaders of the movement called behaviorism decided to let pure reason govern their approach. In a physical universe, one seeks the LaGrangian: the summing-up of present causes in sufficient detail to allow prediction of future effects. Because the universe is lawful and regular, they reasoned, regularities in behavior must be caused by regular influences on the behaving organism. Thus to predict behavior, all we had to do was study the conditions under which it took place with sufficient precision and care. From such studies would come behavioral laws like the laws of physics. Using these laws the psychologist could then not only predict what behaviors would occur, but by manipulating the environment, control behavior.

From the very beginning, therefore, scientific psychology assumed a property of behavior that is precisely the opposite of the one William James noticed. The psychologists decided that if regularities of behavior occurred, they could be traced back to regular antecedents, and that by manipulating those antecedents they could cause the behaviors to occur again. In

Copyright 1988 by William T. Powers. Reprinted from *Conference Workbook for "Texts in Cybernetic Theory,"* American Society for Cybernetics, Felton, California, October 18-23, 1988, 1-32.

253

this way they created an imaginary kind of organism that behaves in a way that real organisms do not behave, and proceeded to spend the next nine decades—so far—trying to make real organisms act like the imaginary one.

This imaginary organism is in fact far older than behaviorism. It came into existence with Galileo and Descartes. The early successes of the physical sciences were based on the fact that in at least some regards, the non-living natural world behaves regularly when subjected to regular influences. The world is a mechanism, and mechanisms do only what they are made to do by outside forces. All of the sciences of life, as they firmed up, sought to apply the same successful methods to determining the mechanisms of life. Behaviorism was born of these earlier approaches; in fact it was directly shaped by the thinking of biologists.

To speak of the "mechanisms" of life is to make a number of subtle but powerful assertions. The subtlest is this: if organisms are mechanisms, they are operated by the world around them. To explain their behavior, therefore, we need look only at their surroundings, and of course at their physical makeup. The physical makeup, however, only establishes the physical thing on which the environment works: without some external force to act on it, the mechanism will do nothing. Whatever it does do, it is caused to do.

This conception of life meant, of course, that to explain behavior we needn't refer to anything inside the organism. No concept of consciousness, thought, or will was needed, because if all behavior could be explained by referring to visible causes, what more could we add to the explanation by assuming inner causes as well? What would be left for them to cause? This line of argument, of course, assumed something that was very far from accomplished: that we could, in fact, account for behavior in terms of external causes.

As the twentieth century got under way, and as more and more scientists pledged alliegance to the principle of external causation, a disinterested observer might have noticed a peculiar fact. Every single attempt to explain behavior in terms of external causation failed. Each one failed, that is, in any terms a physicist or an engineer might apply. Instead of regular responses to outside stimuli, experimentalists kept finding only irregular responses, so irregular that it often took hundreds of trials or hundreds of experimental subjects to reveal that some regularity might lurk beneath the otherwise random-looking data. By the 1930s it had become obvious that the regularities of behavior were all but hidden because of a new property that was named "variability."

So the sciences of behavior became mostly ways of applying statistics to ferret out suggestions of regularity. If there had not been such an enormous commitment to the causal picture of behavior, and so many earnest efforts to show that it was really correct, there would have come a time when these scientists would have stood back, assessed the situation, and given up the basic assumption as a failure. Any physical scientist would have done so long before.

By the 1930s the cause-effect assumption was, however, far too well established to be thrown out or even seriously questioned by mainstream scientists. Essentially all scientific work regarding behavior was based on looking for regular causes of regular behaviors—or at least for correlation coefficients that might be taken as hinting at such a relationship. The scientific world had settled on a general picture of the mechanisms of behavior, and while there was continual wrangling about just how this or that cause affected behavior, there was no disagreement about causality itself.

To this point, the concept of mechanism had essentially only one meaning: a sequence of causal links that began with some primary effect and propagated, one link to the next, until it terminated in some observable event. One part of the mechanism affected the next, and so on to the final effect. But on the morning of August 2, 1929, a Bell Laboratories engineer named H.S. Black discovered a principle that brought a new kind of mechanism into view. On that morning, on the way to work, H.S. Black suddenly understood how to analyze negative feedback.

The artificial control system

Black didn't publish his discovery for four years, but it quickly became the foundation for a new approach to the design of physical systems. The basic problem Black had solved was this: given an electronic amplifier that had part of its output connected to subtract from its input, how could this feedback arrangement be stabilized, so it would not "run away"? Obviously, one answer is not to feed back very much of the output: if the feedback effect is very small, nothing untoward will happen. But what if the net amplification factor, tracing completely around the feedback loop, were very large —say, 1000? This would seem to mean, under the old causal analysis, that any small disturbance would be fed back to the same place with 1000 times the amplitude—and the next time around it would have become 1,000,000 times as large, and so on. Black showed how an amplifier with *any* magnitude of "loop gain" could be made stable, provided that the feedback effect opposed the initial disturbance—that the feedback was negative, not positive. The trick Black discovered was how to make the feedback stay negative.

Systems with large amplification and stable negative feedback soon proved to have some fascinating properties. Their behavior seemed almost independent of their physical properties. Even though stabilizing them meant slowing their responses somewhere in the feedback loop, they were capable of far faster and more precise action than systems without feedback. The speed lost through the slowing factors was far more than made up by the fact that very high amplifications could be used.

Black was primarily a telephone systems engineer, looking for ways to build reliable long-lived amplifiers out of imperfect components. But there was another branch of electrical engineering that found a different use for his principles, the branch that eventually came to be known, early in World War II, as control-system engineering. During the 1930s some engineers were looking for ways of substituting automatic machinery for human beings in certain tasks, primarily tasks that took a whole human being's attention full-time just to keep some simple physical variable like steam pressure or airspeed under control. There was nothing in any existing theory of behavior that could explain how a human being managed to accomplish even the simplest of these tasks. Theories of behavior were long on metaphor and qualitative assertions, but very short on instructions for how to build a machine that would behave as organisms were assumed to behave.

An engineer, some engineer once said, is someone who learns what is necessary to get the job done. In this case, what the engineers had to learn was how organisms really work. They solved this problem from scratch, inventing in the process a new kind of machine. Being interested only in the machine, they didn't realize that they had revolutionized the sciences of life.

It is probably no coincidence that these engineers worked primarily with electronic systems. They were accustomed to systems in which there were no moving parts except at the output, systems in which everything interesting took place in the form of changing voltages and currents. An electronics engineer was perfectly happy to point to a circuit chassis and say, "That's the RF signal, and here's where it gets turned into the IF, and here is the detector that turns it into audio, and here is where the music comes out." In fact, all those currents and voltages were just currents and voltages, until they were named and given functional meaning by the engineer. So there is something appropriate about the fact that engineers working with networks of anonymous and essentially identical electronic signals managed to discover how to build machines that imitate, in a rudimentary way, the kinds of behavior that are accomplished by a brain: a brain in which there are no moving parts and everything that happens occurs in the form of networks of anonymous and essentially identical neural signals.

To shorten the story, the engineers eventually discovered that in order to control some physical variable, a control system had to have certain basic parts, connected in the right relationships. First, whatever was to be controlled had to be continuously represented by an electronic analogue signal. If a position of an object was to be controlled, some measuring device had to be attached to the object so that as the object moved from point A to point B, an electrical signal changed from magnitude A to magnitude B. This was the sensor.

Second, not surprisingly, the control system had to be able to affect whatever was to be controlled. An electronic signal inside the system had to be converted, through an effector, into some physical effect that acted on the variable to be controlled. If an object's position was to be controlled, then the effector would be a motor or a pneumatic piston or a solenoid. For the best control, the amount of action had to be essentially proportional to the amount of driving signal, although it was found that this proportionality could be very approximate.

Having thus dealt with the input and output processes, analogous to human senses and human muscles, the engineers then tackled the third problem, the heart of the matter. Exactly HOW did the sensory signal have to affect the output effector to get the result envisioned—control of the external variable?

It's clear that if the sensor indicates that the position—or whatever—is in error, the sensory signal should operate the effector to make the position or whatever change back toward the right state. A positive deviation should lead to an effort having a negative effect on the deviation, and vice versa when negative and positive are interchanged. Negative feedback. The problem was that you can't simply connect the sensor's signal to the effector and get the right result. If you do that, the control system will energetically force the position/whatever toward the state that creates *zero* sensory signal. If all you want is to keep the position/whatever nailed to the low end of its range of variation that will do fine (although a nail would also work), but what if you want to control something around some state other than zero, or around a variable state?

Consider the poor stationary engineer whose job it is to stand with one hand on a valve wheel and keep a steam pressure gauge at a constant reading. He may not even know that the wheel changes the draft in a furnace and varies the boiling rate of water in the pressure vessel. His job is to keep that needle at the right reading, and all he has to know to do this job is that turning the wheel clockwise will raise the reading and turning it counterclockwise will lower it. Or is that all he has to know?

Actually, he has to know one more fact: the right reading. The dial tells him the present pressure, but not the right pressure. If the dial indicates 328 pounds per square inch, that is too much, and he has to turn the valve counterclockwise. If it indicates 326 pounds per square inch, that is too little and he has to turn the valve clockwise. Only if the reading is 327 pounds per square inch is it all right not to turn the wheel. As the factory is putting widely varying demands on the steam supply, the engineer hardly ever gets to leave the wheel alone and think about philosophy.

So how is the control-system engineer to get that "right reading" into the control system? It's just one position of the needle among all the positions the needle might have, and a phone call from the production manager might result in making some other reading the right one, so 327 pounds now calls for turning the wheel right or left. There is clearly a reference reading against which the actual reading is being compared, and that reference reading, to have any effect, must be carried inside the human being's head. So the control-system engineers had to provide a reference signal inside the control system they were building. The reference signal represented the *intended* pressure.

The sensor represents the *state* of whatever is being controlled as a *signal*, a voltage with an analogous magnitude. It makes sense to compare one voltage to another, and that is what was done: the reference signal was also a voltage. In the nick of time, the 6SN7 vacuum tube came along and (in a circuit called a differential amplifier or "long-tailed pair") provided the basis for an electronic comparator that could generate an output voltage that was reliably proportional to the difference between two input voltages. One input voltage was the sensor signal, the other the reference signal. And now the output of the system could be zero when the input was NOT zero. A motor connected to the draft-adjusting valve could stop turning when the error signal coming out of the comparator was zero, which occurred when the sensor voltage was, say, 32.7 volts, just matching the reference voltage of 32.7 volts. The sensor and reference signals, of course, were calibrated so that one volt meant 10 pounds per square inch in this imaginary but generic design. The sensor didn't read the dial: it was the same pressure sensor that made the needle move.

Now if the pressure was too low the motor would turn one way, if it was too high the motor would turn the other way, and if it was "just right"—meaning that the sensor signal matched the reference signal, whatever its setting—the motor would not turn at all. The control-system engineer could then explain to the stationary engineer that his life of drudgery was over, and also that he had lost his job.

Verbal descriptions of the way control systems work are almost certain to be misleading unless critical details are spelled out with care. The sheer mechanics of speaking or writing stretches out the action so it seems that there is a sequence of well-separated events, one following the other. If you were trying to describe how a gun-pointing servomechanism works, you might start out by saying "Suppose I push down on the gun-barrel to create a position error. The error will cause the servo motors to exert a force against the push, the force getting larger as the push gets larger." That seems clear enough, but it's a lie. If you really did this demonstration, you would say "Suppose I push down on the gun-barrel to create an error... wait a minute. It's stuck."

No, it isn't stuck. It's simply a good control system. As you begin to push down, the little deviation in sensed position of the gun-barrel causes the motor to twist the barrel up against your push. The amount of deviation needed to make the counteractive force equal to the push is so small that you can neither see nor feel it. As a result, the gun-barrel feels as rigid as if it were cast in concrete. It creates the appearance of one of those old-fashioned machines that is immovable simply because it weighs 200 tons, but if someone turned off the power the gun-barrel would fall immediately to the deck. Nothing but the effector, the motor's armature suspended on good bearings in a spinning magnetic field, is holding it in place. The motor does this because the control system is exceedingly sensitive to tiny deviations of sensed position away from the reference position. The gun is so well-stabilized that it resists any amount of push you can exert, without a tremor.

The operator of this gun, on the other hand, can easily make it swivel from one position to another just by turning a knob between two fingers. The knob varies the reference signal. When the reference signal changes, the definition of "zero error" changes, and the control system acts instantly to make the sensed position stay in a match with the new definition. If the operator twiddles the knob idly back and forth, the motor and gears may scream and the lights may dim, but the gun-barrel will also twiddle idly back and forth under precise control.

World War II started only six years after Black published the secret of negative feedback, and sophisticated control systems were pointing gun-barrels before the war's end (I learned to troubleshoot and repair control systems during that war). Into the middle of this feverish development came Norbert Wiener, Arturo Rosenblueth, and Julian Bigelow. They were not the only people to see that control systems behaved in some mysterious fashion as if they were alive—even teenaged Electronic Technician's Mates could see that—but they were the only ones with an ingenious name for this phenomenon: cybernetics, from a Greek word for steersmanship.

Cybernetics

In 1948 Norbert Wiener published *Cybernetics: Control and Communication in the Animal and the Machine*. In this book he showed that the organization of a negative-feedback control system was in one-to-one correspondence with the organization of certain neuromuscular "reflex arcs"; he even suggested new ways of looking at purposive or directed behavior as a whole in terms of control theory. This topic interested many others, and soon gave rise to the Macy Conferences, at which gatherings of scientists explored not only control-system theory, but other topics such as information theory, communication, and self-organizing systems.

The next major publication was W. Ross Ashby's *Design for a Brain*, in 1952. Here Ashby took the basic control-theoretic idea and expanded on it in detail. Among other important concepts, Ashby introduced the idea of "ultrastability," a special property that he gave to a multi-control-system model that enabled it to maintain itself as a control system under drastic changes in its surroundings, even in its own circuitry. This was the first clear statement of a model of organisms showing how they could be responsible for their own organization.

Unfortunately, engineers were under-represented in the early ranks of cyberneticists, one primary exception being Bigelow, who considered himself, however, a proponent of information theory. Perhaps if engineering experts on control theory had been called in early in the game, their conventional and practical knowledge of control systems would have completely stifled the inventiveness that kept cybernetics going. But a price was paid for that intellectual freedom.

It was clear to all the early cyberneticists that control systems behaved in ways that were very different from any concept of behavior that had existed until then. Instead of action being the end of a causal chain, it was simply one part of a closed causal circle. The relationship between organism and environment, when organisms were seen as control systems, was no longer one of obedience to external forces. Instead, the organism itself became an active agent in the world, its inner organization being responsible for what it did. The early years of cybernetics were full of the excitement that comes from seeing a familiar phenomenon in a new light. The implications of circular causality were simply enormous. Studying behavior suddenly became far less important than studying the inner organization of the brain: its inner logic, its use of language, its capacity to do something with incoming stimuli beside respond to them in a blind mechanical way. Organisms began to appear autonomous.

All these new concepts followed, however, from a basic new conception of mechanism that few cyberneticists understood. Most of those who attached themselves to this movement were attracted by what seemed a series of exceptionally coherent insights into the nature of behavior, insights that came, apparently, from nowhere, or at least from a few outstandingly ingenious minds. Most of these cyberneticists understood that somewhere in the background was some technological stuff that had gotten the whole thing started, but they were not technologists and weren't very interested in machines. It was this new collection of concepts that caught their attention. So they began to guess about how such systems might be organized so as to behave in this new way.

There is where the price of ignorance started to be paid. In fact the basic principles of operation of closed-loop systems had been worked out in considerable detail before Wiener and his colleagues ever appeared on the scene. Machines that imitated the purposiveness of human behavior had been designed after a careful analysis of how human beings behaved in that same way (although without any intention of explaining human behavior). The mathematics needed to analyze circular causation, based largely on H.S. Black's work, had matured and was in regular use by engineers. The machines whose behavior inspired the birth of cybernetics *were already understood*. There was no need to guess about how these newly-appreciated phenomena came about.

What cybernetics had to add to this picture was not an explanation of closed-loop phenomena, but a creative exploration of the significance of these new principles as they applied to human behavior. In large part, and to the degree possible at the time, this was done. The way was paved for revising some of our most basic notions of what organisms are and what their actions mean. But at the same time, a body of spurious conjecture appeared, produced by people unaware of or uninterested in the existing knowledge about control systems (or else, aware of it in a peripheral way but convinced that its essence could be captured in a few cleverly-stated rules of thumb).

The most unfortunate aspect of the conjectures was that they were all grounded in the old cause-effect conception of behavior; the radical switch of viewpoint actually required was simply too fundamental to be accomplished without basic knowledge of the principles of control. Those principles, never firmly grasped, soon faded from view. The leaders of cybernetics began, without knowing they were doing so, misleading. One person, who later became a president of the American Society for Cybernetics, announced that he had always considered purposive behavior to be adequately modeled by a drop of water sliding down an inclined plane under the guiding influence of gravity. Another famous cyberneticist, summing up what had been learned during the Macy Conferences, announced that no closed-loop system could avoid runaway oscillations if the feedback factor were greater than unity. Still another proposed that the basic principle of regulation amounted to sensing the cause of a disturbance, and converting that information into a precisely-computed compensatory effect on the controlled variable. Many others proclaimed that control was based on sensing errors, as if error could be observed in the outside world. Others said that control amounted to calculating the precise program of action that would correct an error, and then executing it. Many others said that incoming sensory information "guides" behavior, and another very popular notion was that control consists of limit cycles or alternating sequences of error and corrective actions. Every mistake that could be made was made, authoritatively.

While these views missed the main point, some of them nevertheless contained a grain of truth, and served to keep alive the flavor, if not the substance, of control theory. The basic phenomenon of circular causality continued to be recognized, and its implications expanded. Furthermore, the idea that organisms are active agents was crucial in encouraging explorations of brain models, computer models for the most part, and in leading to the development of new philosophical stances, all pertinent to control theory. The weakness at the foundations was not fatal; at least the implications of control theory continued to be recognized, and continued to attract people who saw that this view made more sense than conventional ones, even if they could not defend it.

We now come to the real subject of this outline: the control-system model I am trying to introduce, or rather re-introduce, to cybernetics. It is not easy for cyberneticists to concede that there is something fundamental about their own discipline that they have missed, especially when the one who makes this claim seems to be an outsider. A certain amount of resistance, even hostility, is to be expected, and I assure the reader that I have already accepted it and discounted it. I have to do so, to remain consistent with the principles I believe to apply to human nature.

But something is demanded of cyberneticists, too; they must at least take under advisement the possibility of thinking the unthinkable. I ask no more than understanding of what I propose.

Cybernetic control theory

While I already knew a little about control theory at the time, my lifelong interest in applying it to human behavior began only after I read Wiener and Ashby in 1952. It seemed to me that they had uncovered a vastly important principle of behavior, new to the life sciences. Being unknown and feeling ignorant, I determined to learn more about control theory and its applications to behavior, so that some day I could enter those exalted halls of cybernetics with something to contribute. This project began in 1953, in collaboration with a physicist, R.K. Clark. We were soon joined by a clinical psychologist, R.L. McFarland, and began to learn control theory in depth, my role being that of an engineer/physicist who was designing and building control systems as part of the job of a medical physicist. Clark really made the whole project possible by finding us both a position at the V.A. Research Hospital in Chicago, where I worked as his assistant. McFarland was the Chief Clinical Psychologist there, and made important contributions in translating our somewhat austere models into terms that conventional psychologists might conceivably understand.

Our first paper describing the control-system model was published in 1960, in the shadow of Miller, Galanter, and Pribram's book on the organization of behavior, where the TOTE unit acquired its unfortunate lease on life. I will not bore the reader with tales of the meager acceptance that greeted our publication: cyberneticists have had their own problems, for similar reasons, with the Establishment.

This brief review of my own history is by way of saying that my interest in control theory was originally inspired by cybernetics, and was always intended, at least as a background hope, for use by cyberneticists (as well as psychologists). I thought, for many years, that I was simply catching up.

Neither will I bore the reader by re-running the laborious process by which we arrived at the final model, after backing out of many blind alleys. I will pass over the ensuing years of intermittent discouragement, the regrouping that ended with my book in 1973, my subsequent tentative forays into the American Society for Cybernetics, and the rise of the Control Systems Group, that rumor of Visigoths poised on the borders of cybernetic civilization ready to plunder and rape and otherwise violate the comfortable ways of the ASC. None of these matters will be important if the basic concepts of this theory are clearly understood. We have all been through the wars. We are all on the same side. Let's get to it.

The nature of control

The first thing that must be understood is that control is something that a control system does, not something that is done to it. The second thing is that in a control system there is no "controller." Control is a phenomenon that arises when an active system, constructed in a specific way, interacts with its immediate environment. The third thing is that the relationship between control system and environment is not symmetrical. Even though each affects the other, only the control system controls. The word "environment" means here the passive physical environment that takes no action of its own, but behaves as it is made to behave by natural forces: the world of the physicist. The presence of other control systems is a complication we will take up later.

A control system senses its environment and acts on it. Sensing means representing, and representing, if it is to mean anything reasonable, means analogizing. A sensor responds to some specific aspect of its environment, some variable outside the sensor, by generating a signal that is a quantitative analogue of the state of the variable. Bear with me for now: this concept of representation will become more interesting.

Acting means generating some physical effect whose magnitude and direction depend smoothly on the magnitude and sign of a driving signal inside the control system. Again, bear with me: we are speaking of the foundations of more complex actions.

As explained earlier, the sensor signal representing the external variable is compared with an internal reference signal that is of the same physical nature as the sensor signal. The result is an error signal that is zero only when the sensor signal matches the reference signal.

The action of the system is driven by the error signal.

In order for control to appear, the parts of this system must act in specific ways. The sensor signal, for example, must vary over a range from minimum to maximum as the external variable goes through its whole possible range of change. This relationship establishes the range within which control is possible.

The action of the system must affect the external variable at least in the dimension that is sensed. If an action caused by a positive error signal changes the sensed variable in one direction, the action caused by a negative error signal must change the variable—as sensed—in the opposite direction.

The overall effect of these relationships must be that the action driven by either sign of error signal must tend to alter the external variable in the direction that makes the sensor signal come closer in magnitude to the reference signal, so that the error signal becomes smaller. This is the basic requirement for negative feedback.

These requirements give us the qualitative basis for control phenomena. But there is a critically important quantitative basis as well, which accounts for the asymmetry of control.

The error signal drives the output action. It makes a great deal of difference how much error is required to produce a given amount of action. The ratio of action to error is called the *error sensitivity* of the control system. The output function, the effector of the control system, not only converts from signal units to physical-world units of effect, but it enormously increases the level of energy that is involved in all variations. The output function is a transducer, but it is also an *amplifier*.

The output action of the system is connected to the external variable through an environmental link. In this link the laws of thermodynamics prevail: no more comes out than went in. Between the action and the effect on the external variable there is usually some degree of *loss* of effect. There may be a change in energy level in passing from the external variable to its sensory representation, but if we normalize both variables to their total range of change, there is no amplification. Almost all of the amplification (that is not simply a change of units) that occurs in this control process occurs in the output function, in the conversion from error to action. Here thermodynamics means nothing: the system is supplied from outside with whatever amount of energy it expends. The books do not have to balance: this is a thermodynamically open system.

It is a peculiarity of control systems that causation often seems to reverse itself. If we compare two control systems with greatly different error sensitivities, our first guess might be that the system with the greater error sensitivity, all else being equal, would produce the greater amount of action. What actually happens is that the system with the greater error sensitivity contains the smaller error signal, and its action is essentially the same as what the other system produces. If you double the error sensitivity, the result is very nearly to halve the error signal, not to double the amount of action.

There is one last consideration that has nothing to do with the process of control itself, but which is one of the major reasons why control is necessary: disturbances. The external variable is affected not only by the system's action, but by the world in general. The temperature of a house is affected not only by the furnace's output, but by heat entering, leaving, or being generated by other sources in the building. The path of a car is affected not only by the driver's steering efforts, but by crosswinds, tilts and bumps in the road, soft tires, and misalignment of the wheels. A savings-account balance is affected not only by depositing and withdrawing money, but by service charges, computer errors, and crooked employees. Variables that organisms control are controlled because they will not spontaneously come to the states desired by the organisms, and even when brought to those states, will not stay there.

The physical environment is in a continuous state of variation, so much so that no specific action can have just one specific consequence. There can be no such thing as computing an action that will have a desired result, unless one has taken great pains to shield those results against all normal independent influences. That may be approximately possible in the laboratory, but it does not happen in normal environments.

Furthermore, as we are beginning to hear, the lawfulness of the physical world itself is largely illusory even discounting Heisenberg. Many natural phenomena are so sensitive to slight variations in initial conditions that even though we can prove, by backward reasoning, that they are lawful, we cannot establish initial conditions accurately enough to turn those deductions into reliable predictions. The behavior of higher organisms is clearly one of these phenomena. Behavior results from the application of muscle forces—not very reproducible in themselves—to the masses of the body. The result is not "movement" but acceleration. Even to turn an effort into a position requires a double time-integration, which vastly magnifies all force variations, and by greater and greater amounts as time progresses. And this does not begin to take into account the indirect effects of limb movements that, in order to produce the larger patterns of behavior, must be integrated again and again, all the while being subject to unpredictable disturbances. It is not necessary to invoke control theory to show that the old causal model of behavior is wrong: all we need do is look realistically at what is involved in making "the same behavior" occur twice in a row in a disturbance-prone and semi-chaotic universe.

If organisms simply behaved blindly, the consequences of their actions would be essentially unpredictable. The same action applied ten times in a row would have ten different consequences, in most cases radically different. The physical world, uncontrolled, drifts in a kind of gigantic Brownian movement, showing order on an intermediate time-scale but for the most part simply changing aimlessly. Control systems impose order on this aimless drift. The automobile, buffeted by winds, jolted by bumps, dragged by uneven friction, wearing out asymmetrically from one minute to the next, nevertheless clings to a path that deviates by no more than one or two feet from the right path in 100 miles. This regularity is wholly unnatural, and can be accounted for only by knowing that there is a control system at the steering wheel.

The fact that there is behavior at all shows us that there is control.

To grasp the behavior of a control system correctly, it is necessary to think of all parts of the system at once. Control is not a sequential process, but a process of continuously and energetically maintained equilibrium among all parts of the system and between the system and external influences. If a disturbance arises that tends to change the external variable being controlled, the system does not wait to act until the disturbance has finished its work. Instead, the action of the system begins to change the instant there is any deviation of the sensor signal from the reference signal. Because this action opposes the error, it also opposes the effect of the disturbance. As the disturbance increases and decreases, so does the action opposing it increase and decrease. The sensor signal, in this process, varies slightly away from the reference setting, but if the error sensitivity is reasonably high only a tiny amount of error is needed to keep the action balanced against the disturbance at all times. For all practical purposes the action prevents the disturbance from affecting the controlled variable.

You will notice that some familiar concepts customarily associated with control processes are missing here. The first missing factor is any ability of the control system to sense the *cause* of a disturbance of the external controlled variable. While a more complex system could sense the cause of the disturbance, doing so would not materially improve control. The control system responds only to deviations of its own sensor signal from the reference signal. Why there is a deviation, whether it is due to a single cause of disturbance or to the combined effects of a thousand independent causes all acting at once (the normal case), is irrelevant. All the control system needs to monitor is the controlled variable itself: if the controlled variable starts to depart from its correct state, the system acts directly on it to keep it where it belongs. There may be a few circumstances in which "feed-forward" would be advantageous, but it can never substitute for the basic process of control. I should add that *pure* compensation, in which only the state of the disturbance (not the controlled variable) is sensed and a compensating action is calculated and applied along with the effect of the disturbance, will not work at all in most circumstances. It may seem to work on paper, where we can represent variables by simple whole numbers and give the imaginary system knowledge of all disturbances acting (and of the links from each disturbance to the controlled variable), but in the real world it can't even come close to explaining what we observe.

Another missing factor is any provision inside the control system for computing the proper amount of output to correct a given error. The only thing approximating an output computation is the amplification of the error signal, the system's error sensitivity. In order to compute the right amount of output to produce a given effect on the controlled variable, the control system would need a great deal of information that its simple sensor signal does not carry. It would need to know the momentary properties of the physical link connecting its action to the controlled variable, and it would need to know what amount and direction of disturbance will be acting at the time when the output calculation is put into effect. To get the required information it would need a vast array of extra sensors and a very large computer programmed with the laws of physics—and the ability to predict future disturbances. Furthermore, it would need to know about its own properties, because the instant that the output computation began to have its effect, the input variable would change to a different state, making the computation obsolete. The concept of "computing the appropriate action" is not only superfluous, but amounts to a very poor design. In the real world, human beings often try to control complex events in this way, thinking that logically it has to work, but in fact such efforts usually prove fruitless, as witness the attempts of the Federal Reserve to regulate the economy by diddling interest rates.

Finally, also missing is the entire concept of a "controlled action." Control systems do not control their actions: they vary them. What they do control is the variable affected both by the action and by disturbances. And in the final analysis, what they *really* control is the sensor signal that represents the external variable. All the rest of the system functions to maintain the sensor signal in a match with the reference signal. The action of the system is determined at every moment by the nature of the feedback link to the controlled variable and by the amount and direction of net disturbance that is acting. If the action itself were controlled, the variable could not be stabilized against disturbance. If the driver of a car controlled the *steering wheel* instead of the position of the car, the car would go immediately into the ditch, because no one position of the steering wheel will keep the car on the road for very long.

Fig. 1 shows the basic relationships we have been talking about.

A hierarchy of control

What we have seen so far would probably be called a "homeostatic" system. We have a system that maintains a one-dimensional variable at a constant level matching a fixed

272 *Living Control Systems*

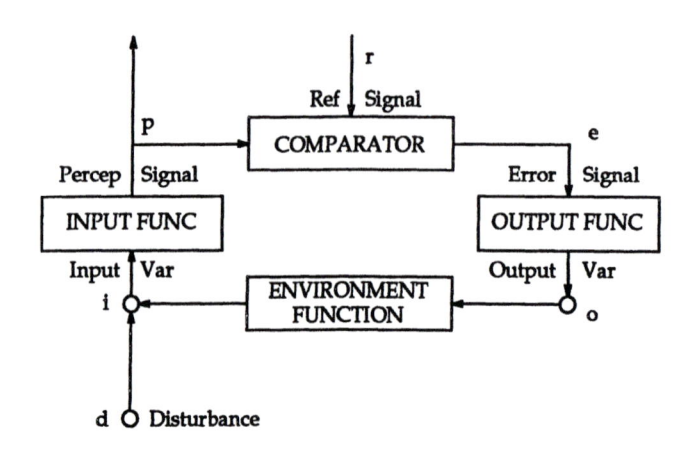

```
                          r
                     Ref ↓ Signal
           P       ┌──────────────┐
    ┌──────────────│  COMPARATOR  │         e
Percep│ Signal     └──────────────┘    Error│ Signal
┌──────────────┐                      ┌──────────────┐
│  INPUT FUNC  │                      │ OUTPUT FUNC  │
└──────────────┘                      └──────────────┘
  Input │ Var        ┌────────────┐    Output │ Var
   i    ◯────────────│ENVIRONMENT │───────────◯ o
        │            │  FUNCTION  │
        │            └────────────┘
        │
    d  ◯ Disturbance
```

Figure 1. Generic control-system diagram

reference signal. This system might behave very energetically as disturbances come and go, but the net result of its action would be a variable that is held constant.

By now, however, it should be clear that the control system's action focuses on maintaining its own sensor signal in a match with the reference signal. Nothing was said that specifies the setting of that reference signal, and nothing was said to limit the reference signal to a single fixed value.

If the reference signal varies in magnitude, the first effect will be to *create* error. Instead of the sensor signal departing from the reference signal, the reference signal departs from the sensor signal, but the result is precisely the same: an error signal that is highly amplified to produce action. The basic arrangement has not changed: the system will still be organized to alter the sensor signal in the direction that makes the error smaller. But now its action will have the effect of making the sensor signal change, rather than holding it constant.

In a well-designed control system, errors are never allowed to get very large. Consequently, when the reference signal changes, the output action will drive the controlled variable to change right along with the reference signal. This is the gun operator twiddling the control knob. Changing the reference signal is a way of changing the external controlled variable in a predetermined way—namely, the way the reference signal changes. If the reference signal changes smoothly from a low value to a high value, so will the controlled variable change, quite without regard to any other physical influences acting on it. The control loop will automatically produce whatever fluctuations in action are required to make the controlled variable obey the reference signal rather than other influences.

So whatever is capable of manipulating the reference signal is also capable of manipulating a variable in the environment of the control system. The way that variable changes is determined by the cause of the reference-signal changes, and more important, ceases to be dependent on all the physical laws that would otherwise determine how it behaves. The control system has taken over that variable, cut it out of the normal flow of inanimate nature and made it behave as the control system—or as the manipulator of the reference signal—wishes it to behave. The aimless drift that the variable would naturally exhibit is replaced by purposive change. Regularity has been imposed on Chaos.

Note that we still do not have purposive *action*. The actions of the system are still dictated by disturbances and by natural resistance of the variable to being changed. For any given state of the controlled variable, the action might be found anywhere within its possible range, depending on what else is doing something to the controlled variable, or trying to. Purpose can be seen only in the controlled variable itself—in its variations that have been rendered immune to the normal forces affecting it. The purposiveness of a home thermostat is not to be seen in the furnace's turning on and off. It is to be seen in the steady temperature of the room where the sensor is located: 68 degrees in the daytime, and 62 degrees at night, when the little purposive computer lowers the reference signal for the temperature-control system. Rain or shine, summer or winter, the temperature stays at one or the other intended level. The furnace turns on and off as it must. Controlled variables, not actions, contain the evidence of purpose.

In the human body, at the lowest level of behavioral organization, there are something like 600 to 800 small control systems, each of which controls the sensed amount of strain in one tendon. The signal representing tendon strain is sent to the spinal cord, where it is compared (by subtraction) against a reference signal arriving from higher centers. The resulting error signal drives the muscle associated with the same tendon. These systems are small, but they are not weak: the range of strain that can be detected and controlled ranges from about a tenth of a gram up to something over 300 kilograms, in the system associated with a normal biceps muscle.

The reference signal that reaches the spinal comparator has been described regularly as a "command" signal, its function being to cause a specific amount of muscle contraction. But that is not how it works. The reference signal specifies how much signal is to be generated by the sensors that detect tendon strain. If disturbances alter that strain, the local control loop will automatically raise or lower the muscle tension to leave the net strain the same. It is the stretching of the tendon, not the contraction of the muscle, that is under control.

More specifically, it is the *signal* analogous to tendon strain that is controlled. In each case, this signal follows a branching path. One branch goes to the spinal comparator, as mentioned. The other branch continues upward, or inward, carrying a copy of the sensor signal in the direction from which the reference signal is coming. When everything is working properly, the upgoing copy of the sensor signal varies exactly as the descending reference signal varies. From the standpoint of the higher systems generating the reference signal, the effect is to control a sensation of effort simply by varying a signal standing for the amount of intended effort. The brain "wills" an effort by emitting a reference signal: immediately, that same amount of effort is experienced. The lag is imperceptible, amounting to no more than 20 milliseconds. It's no wonder that we have trouble separating the sense of willing an action from the experience of the action occurring. Paralysis, of course, makes the difference frighteningly obvious.

We have now created a class of control systems, the set of all effort-control systems. Everything that a human being does that could be called overt behavior is done by varying the reference signals reaching these systems. *Everything.* Whether a person is playing a piano concerto, painting the Mona Lisa, pressing the button that starts a war, making a lying speech to skeptical constituents, skating for an Olympic medal, or pounding on the keys of a word-processor, the acts involved are all accomplished by varying the reference signals reaching these 600 to 800 first-order control systems.

No system higher than the first order can act directly on the environment by generating physical forces. The actions of all higher systems consist entirely of generating outgoing neural signals. There are no moving parts in this system above the first level. There are only signals, and systems that receive, manipulate, and generate signals.

This is not the place to present 30 years of elaboration on this concept of levels of control. I will only try to sketch in the basic relationships that seem reasonable to propose. As far as I know, there is considerable neurological evidence in support of these suppositions, and nothing known to speak against them. But I am not pretending to be a brain researcher: I'm only trying to put together a feasible picture of an organization that has, within the bounds of what we know, a chance of actually existing. Perhaps these suggestions will raise some questions in the minds of real brain researchers. I'm far from the first to suspect control systems in the brain, but I don't believe that anyone else has approached the problem quite in this way (at least before I did). My little claim to fleeting fame.

Having isolated the first-order behavioral control systems, we now have a collection of incoming sensory signals, a subset of which is under control, and a collection of outgoing signals that become reference signals for the first-order systems. We can ignore the probability of cross-connections and other complications at this and other levels, in the interest of seeing the big picture first.

It's clearly possible now to think of a second level of control. At this second level, a control system would receive some set of first-order sensory signals (most of which come from receptors not involved in effort control), and would re-represent this set of signals by combining them in perceptual computing functions to create a new set of signals. Each second-order perceptual signal thus produced will represent some new type of invariant of the first-order world (every single-valued function of multiple variables generates some sort of invariant). I have reason to think, but will swallow the temptation to elaborate, that each new level actually represents a new type of variable in exactly the sense of Russell's Theory of Types.

Once an aspect of the first-order world has been represented as a one-dimensional second-order perceptual signal, we can quickly assemble a control system. We need a reference signal from still higher up, and a comparator to generate an error signal. And we need an output function that will amplify the error signal and send the result in the form of reference signals to *all the first-order control systems that can affect the second-order perceptual signal*. The effect may be direct, through pathways inside the body, or indirect, through pathways that include the external world. The effect may be achieved through altering the external world, or by altering the relationship of parts of the body to it, as when the eyes move.

How many second-order control systems might exist? A great many: a better question would be, how many can be active at the same time? Here there is a fundamental limit. The number of first-order control systems sets one limit on how many independent combinations of muscle tension can be produced at the same time. The number is large, but it is not infinite.

A second limit exists, set by the number of different functions, independent of each other, that are perceptually computed from the set of all first-order perceptual signals (at any one time). At most, 600 to 800 such signals might conceivably coexist, but in fact the likelihood of that many independent functions being discovered by the brain has to be very small. Let us just say that there is some number of independent dimensions of the first-order world that could be simultaneously computed, and that it must be considerably less than 600.

Why is this limit on numbers important? Because of a consideration left out of the discussion so far. Even just on anatomical evidence, we know that each spinal comparator neuron receives not just one reference signal, but in most cases hundreds of them. There can be only one net reference signal at a time for one first-order control system, but because the converging reference signals can have both positive and negative effects on the net setting, this net reference signal has to be considered as the weighted sum of the outputs of many higher-order systems. We can say "second-order" systems; there are arguments against reference signals skipping orders on the way down in a control hierarchy (such signals would be treated as disturbances and canceled).

We thus have a picture in which some number of independently-acting second-order control systems act by sending multiple amplified copies of their error signals to many first-order control systems, specifically those whose actions can alter the second-order perceptual signal directly or indirectly. The second-order systems therefore share the use of overlapping sets of first-order systems. No one second-order system determines the net reference setting for any one first-order control system. The net reference setting for one first-order system is always a compromise among the demands of all the second-order systems that affect it.

What's interesting about this arrangement is that it can actually work. The crucial part of this sharing of control is not the separation of output effects—those are simply added together, with the appropriate sign to maintain negative feedback around each loop. What matters is that all the second-order input functions produce perceptual signals capable of independent variation: the *input* functions must be linearly independent. Given these conditions, we have a well-known setup for the solution of large sets of simultaneous equations by analogue computation. Digital computers can be set up to do the same thing, far more slowly, using "methods of steep descent" and other arcanities. It is possible for many second-order control systems to maintain quite independent control of their own perceptual signals, despite having to act through a set of shared first-order control systems.

Fig. 2, thought up and drawn by Mary Powers and a handy program, shows a few of the arrangements possible in a large hierarchy of control systems. Of course only a few connections are shown, with some deliberately confined to input or output effects for clarity. In the middle and on the right are shown some short-circuit connections, in which the outgoing reference signals bend back to become inputs to the same systems, without involving lower systems or the environment. This is the "imagination connection" that enables us to think—to envision the effect of doing things but without doing them.

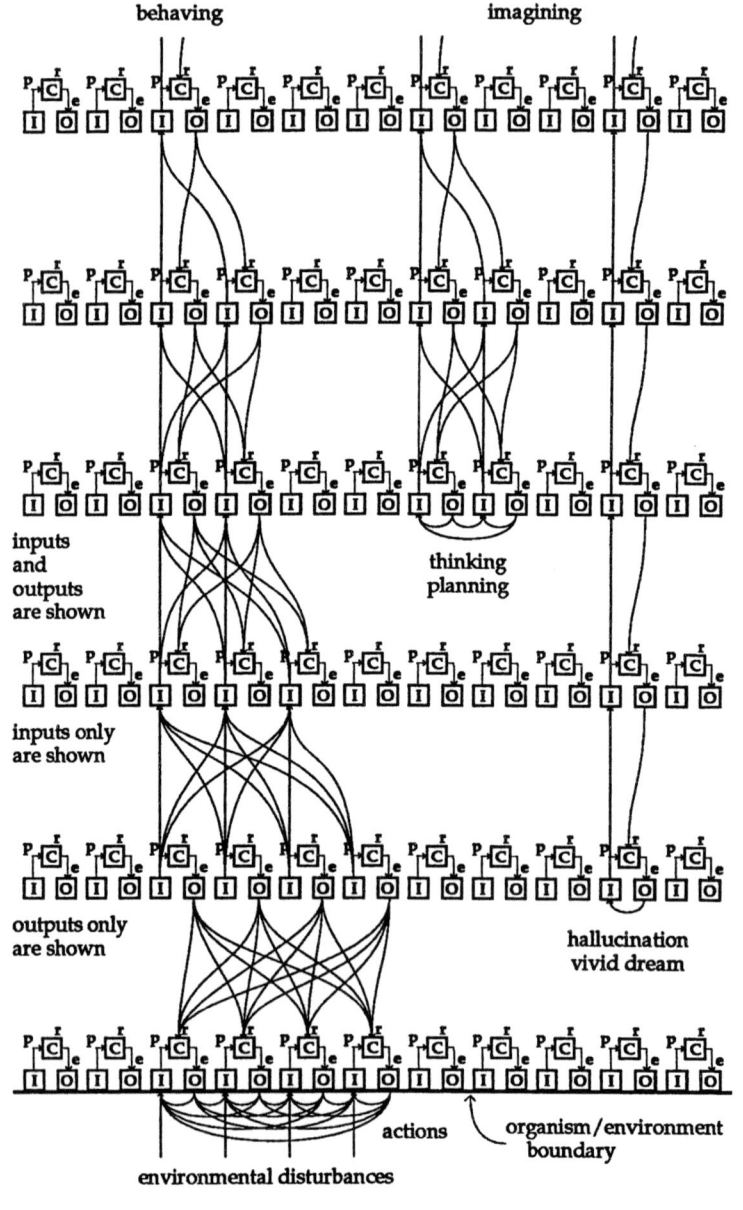

Figure 2. A hierarchy of control

Above the level of the imagination connection, the perceptions are perfectly normal, except perhaps for the combinations in which they can occur. We have a sense "that" something is happening, without the lowest-level details to make it vivid or real.

This diagram has a vague resemblance to a real nervous system, which would become much stronger if at each level we stretched the connecting lines and clumped all the input functions together, and all the comparators and output functions together. Then we would have a realistic picture of the sensory nuclei, the motor nuclei, the upgoing and downgoing tracts, and the collaterals that run crosswise at every level in the brain.

At every level that may exist, we can expect the same sort of arrangement. Each new level of perception creates a new class of entities that can be controlled by varying reference signals at the next lower level. If you trace out any higher-order control system, you will see that the control loop *always* (except for imagination) involves effects in the external world. This permits us, as external observers organized in the same way, to discover the aspects of the shared world that are under control by another organism, even though those aspects be highly abstract. All that is required is that we learn to apply the same stages (or equivalent stages) of perceptual computation to the basic sensory input we are getting—from what we presume to be a common environment. This is how we attack the problem of communication under the control-system model.

There are obvious questions about the highest level of control, and obvious answers that I will not spend time on here. I hope it is suspected that far more could be said about this hierarchy than I have said. Most people take about two years to get the full picture of this model even when they're trying; we won't get that far in one paper.

There are two main subjects that still really need discussion —I will abandon the notion of getting into the biochemical control systems and evolution, because this is already a very dense and long presentation. One subject is epistemology, which takes on a particularly important significance in this model, and the other is reorganization, the key to the development of an adult control hierarchy and also, although I won't go this far, the route to understanding physiological growth and the evolution of species. I want to show how the control model bears on two subjects that have become centrally important in cybernetics over the past ten or fifteen years. And I would like to say at least a word or two, at the end, about the picture of human existence and aspirations that control theory can give us.

The view from inside

To this point we have been looking at control systems and the world with which they interact from a viewpoint that is convenient but artificial. From where we stand, or float, we can see the physical environment surrounding the body, the brain and nervous system inside the body, and the signals spreading through millions of channels in the brain. Our X-ray eyes penetrate the skin to reveal muscles contracting and relaxing, putting stresses on tendons that give way slightly, exciting the little sensory nerves embedded in them. In the outside world we can see objects, but also the forces and influences that connected them together. When we put matters that way, it has to be clear that this entire picture is imaginary. It is, in short, a model: a model of a brain in a model of a world.

Here is a simple question: according to this model, where is the model? If you look at Fig. 2, you will see those imagination connections that allow higher systems in the brain to generate perceptual signals for themselves without causing them in the normal way by acting on the external world. The model says that this imaginary picture of the brain and the external world exists in the brain, and is created inside the brain. My brain. Perhaps, a little bit, your brain too.

In particular, the model implies that all these things we experience, whether in imagination or "really" (there is no remarkable difference), reside in the upgoing perceptual pathways. This leads me to make a proposal for which there can be, in the nature of things, no direct evidence, but that does make a lot of things fall into place rather neatly. It is this: the objects of experience of any kind exist in the form of perceptual signals continually rising through the brain.

This proposal in no way pins down who, what, or where the perceiver is, the noticer, the observer. It concerns only *that which is observed*. The *objects of observation*, I am proposing, are neural perceptual signals in the brain.

If you were to spend a few decades systematically and skeptically examining the real solid three-dimensional physical world that you see, feel, hear, touch, and taste, I claim that you would find it to consist of a number of types of experience. From simplest to most complex, I claim that these types can be named roughly this way: intensity, sensation, configuration, transition, event, relationship, category, sequence, program, principle, and system. The words need some elaboration to make their intended meanings clear, but you get the flavor.

These types of experience have an interesting relationship to each other. The ones farther along in the list—"higher"—depend for their existence on the existence of types lower in the list (I do many things backward: my list goes from bottom to top, and I write it left to right). Furthermore, if you want to change a particular experience of a given type, you will find it necessary to change experiences of lower types. Those relationships, however, are not reciprocal: a lower type of experience does not depend on a higher one, and can be changed without changing a higher one. As we go up the list, the relationships between types are the relationships between successive stages of invariants, each stage abstracted from the previous one by a new rule, as in Russell's Theory of Types.

This, not by accident, is exactly the structure of the perceptual part of the control hierarchy in Fig. 2. It also seems to be the structure of the perceiving functions at various levels in the brain, give or take some topological transformations, and allowing for the fact that models are always neater than nature.

But this is not just a structure of perceptual functions: it is a structure of control systems. A control system at any level acts on a world consisting of lower-level control systems, the means of acting being to send varying reference signals to some of the lower systems. These control actions ultimately result in the lowest-level systems doing things to the outside world, and thus to the lowest level of perceptual input signals, the intensity signals. The first-order signals are abstracted to become second-order signals, and so on until we reach the system we began with, where the effect of that system's action is to maintain its own perceptual signal in a match with the reference signal it is being given from above.

But here we are floating in space again, while the point, if I haven't mentioned it, is to see how it is to *be* a system like the one in Fig. 2.

When you *are* a system like this, you find that by acting you can alter the world you perceive. When you learn its rules well enough, you can learn how to make many of those perceptions come to states you have experienced before and liked, or to stay away from states you have experienced before and didn't like. When you see a flower, you can move it to your nose or your nose to it, use your diaphragm to pull air in, and experience a scent that you judge as pleasant. If it's a pretty flower that ought to have a scent but doesn't, you can supply a scent in imagination. You can supply a scent at a low level, like a hallucination, or at a higher level, like an impression of niceness.

Whatever you do alters your perceptions: that's how you know you're doing something. You perceive your own efforts and their immediate effects such as skin pressure; you perceive effects of those efforts as objects change their (visible) positions, orientations, and velocities. You use your ability to control your limbs as ways of controlling other objects; you use control of other objects to create movements and events in relationship to other movements and events; you control movements and events to maintain certain categories of experience in the states you intend; you maintain these categories in sequences that constitute progressions of familiar kinds; you adjust these progressions according to rational decisions, choices, tests, and symbolic equivalences; you carry out rational processes in support of general principles that you defend, and you maintain those principles as a way of sustaining whole systems within which you live and experience and which you try to maintain, systems like a self, a science, a society, a culture, a world, a universe.

All of our actions, according to this control model, are part of a process of controlling perceptions. To understand this idea properly, you have to abandon all the meanings the word control has accumulated, meanings that represent, mostly, bad guesses as to what is going on. Controlling does not feel like trying: it is lack of control that feels like trying. Controlling is just *doing*. You don't have to "try" to look at something—you just look. Your oculomotor control systems snap the object you want to look at to the center of your visual field, and there is no sense of trying at all. You don't "try" to write your name; you just write it. By far the majority of control processes that go on at all these levels are skillful, swift as thought, stable, and seemingly effortless. You form an idea of what is to happen and it happens at the same time. You just do it. There is no process of laboriously selecting some intended perception, figuring out a way to get closer to it, and then painfully working your way toward zero error. That only happens when you don't know what you're doing. Mostly our perceptions track our intentions with no perceptible lag. That's why, sometimes, they're hard to tell apart.

Of course at the higher levels of control, particularly the cognitive levels, things happen more slowly unless we're imagining. There has to be time for all the lower-level systems to bring their perceptions to whatever the momentary net reference signal specifies. The lowest level systems have a lag of perhaps 50 milliseconds, whereas the highest ones, operating at their fastest, may lag as much as half a second or a second. Some control processes may take much longer than that: I'm involved in one that has been going on for—let's see, 1988 minus 1953 plus one—thirty-six years. Of course a wise person doesn't tolerate protracted error; he or she redefines the controlled quantity so it can in fact be held at its reference level without large error. I'm making progress, that's more like it.

To say that behavior exists in order to control perceptions is not to say that all perceptions are under control. Much that we can see happening around us happens without benefit of our advice or effort. But we do come to "expect" the world to be a certain way; that is, even without specifically intending to do so, we set up reference signals against which we compare perceptions even when we have no direct way of affecting them: an inner model of how the world should be. As long as the world matches these expectations, we experience no error and go about our affairs normally. But just let the sun rise in the

West one morning, and see how much error you would experience, and how frantically you would start to act to try to do something about this gross mistake.

You can see that this model implies an epistemology. If what we experience consists entirely of perceptual signals in the brain, it follows that we do *not* experience the causes of these signals. The causes lie outside, according to this model, beyond our sensory endings where we, the observers who experience only perceptual signals, have no contact. Our motor efforts disappear into that world, and we know nothing of what they do to the world until the effects return to cause changes in our intensity signals. What we can know of that external world consists only of what we can sense, and what we can imagine. Sensing and imagining occur inside, not outside, the brain.

How would a brain organized as this model is organized ever know that an external world, other than the apparent one, actually exists? There are at least two kinds of evidence available. One kind is the fact that in order to bring any perception under control, the brain must discover how to manipulate reference signals to have the required effects. This can be done only by trial and error, with perhaps a smidgin of genetic help. The relationship between what must be done and the result that it has constitutes a model of some "property" of the external world. The fact that stable properties can be found is evidence that there is something lawful and stable outside the boundaries of experience. In more formal surroundings, this is called "scientific method" (except in the behavioral sciences, where scientific method means assuming a cause-effect model and then throwing out all data that doesn't conform to it).

The second line of evidence is found in the very fact that control is necessary. The world will not usually meet our needs, desires, or expectations unless we do something to it, and even when we have learned how to maintain the world as we wish it to be (in the respects we can affect) we find that we still have to vary our actions in order to maintain it in any particular state. In other words, those perceptions are subject to influences other than our own actions. Disturbances. The driver of a car can deduce the direction and strength of a crosswind that he cannot sense in any way, simply by observing how he is holding the steering wheel. He quite automatically varies the position of the steering wheel in the way required to keep the scene in the windshield constant, showing that the car is in the right position on the road. He has no preference for wheel position. Thus he can "see" the crosswind, deduce it from his own control actions, without any other way of sensing it. He could, of course, be wrong: there could be something horribly wrong with the car.

That's really a third line of evidence: we can be wrong. We can go through half a lifetime or more thinking we have really gotten something nailed down, have full control and a competent model of what is happening, only to have some trifling incident turn our whole idea upside down, utterly destroying, for a while, our confidence in our ability to know anything. Such an experience, however, should give us more confidence, not less. What should make us lose confidence is finding that we can no longer detect the mistakes that tell us we can still, somehow, be in contact with reality.

This is certainly not a philosopher's approach to epistemology; it's a purely practical approach. I think that practicality, pragmatism of the right sort, is the key idea here. Knowing that it's all perception, we will think in new ways about most of our own experiences and actions. But will we then give up making models, just because we know they are "only" models? That would be foolish, because then we would be giving up the basis for giving up models, wouldn't we? I think the best course is to admit that what we call knowledge consists entirely of models, models of a body, of a brain, of a physical and chemical reality, of a society, of everything. Rather than giving up models, we should become conscious of the process of making models.

If we know we're making models, we won't go around telling people that they are wrong for trying out different models, or that they are right even if their models are sloppily constructed and unconnected with any other models. We should be looking to make all of our models consistent with each other, and worrying very seriously when they are not, and being fussy about what we will accept as a model. We ought to test the hell out of our models, because if they don't behave the way our experiences behave, they are worse than sloppy: they're delusory. They're useless. They're dangerous.

Of course when we know we're making models, we can be free to try out any ideas we please, as long as we realize that we're playing what is in the end a serious game. We are trapped in here, folks, and our very survival depends on making models that in some way reflect the regularities of the real universe that is right out there, but that we can know only approximately and only by way of models. Fun and games make life interesting, but somebody has to take out the garbage. But it's not that bad. Making models is really *fun*. Hello?

One last point before we leave this subject barely touched. I have made the claim that our experiences of the world fall into eleven types (more or less). Does this mean that the real universe is organized that way? I think my answer would be pretty obvious: of course not, although we can conjecture that there is some reason for these particular types to have evolved (the evolution-model). Basically, the types of perceptions are determined completely by the types of perceptual functions that are applied to each level of signal, and it is highly probable that each person organizes perceptions, within each type, quite differently. But there is a miracle going on that anyone interested in epistemology should acknowledge.

The miracle is that we can talk together at all about anything. Everything enters our nervous systems at the lowest level, becoming available to the brain first as a huge collection of identical intensity-signals. It takes many layers of information-processing before those intensities can be turned into the perception of a sentence, and more yet before a string of grammatically and syntactically ordered words can be used to evoke a non-verbal experience, the perceptual meaning of the sentence. I must turn my meaning into a sentence, and utter it, and you must turn the sound-intensities back into a sentence and the sentence into a meaning before anything resembling what is in my awareness springs up in your awareness. So how do we ever come to believe that the meaning you get is the one I intended?

Very often it's not the same. We only think it's the same, and sometimes fatally, assume it's the same. Finding out if it *is* the same is basically impossible, but even reaching some level of confidence in the sameness requires a long process of back-and-forthing, of cross-checking, of "If I understand you correctly, then when I do *this*, you'll do *that*."

Yet, look at this: over ten thousand words so far, and I still have some hope that you are with me. What you have made out of all these words, I will know only when you do or say something relevant to them—as I intend their meanings. Epistemology is a very faint echo of the real problem, which is communication.

With that, let us pass on to the final topic.

Reorganization

I'm going to give short shrift to this subject partly because my endurance in sustaining this long narrative is beginning to wear down about as far as the reader's must be. This is a critically important subject; unfortunately, I don't know much about it, and can speak only in generalities. This is one place where I really wish I were a good mathematician.

The idea of reorganization is an essential part of this model, and has been since its beginnings. It was suggested—laid out pretty completely—by W. Ross Ashby in his notion of "ultrastability," and independently by Donald T. Campbell as "blind variation and selective retention." The basic idea is simple, and older than either Ashby or Campbell.

There are many forms of learning, but the most fundamental is learning something for which there is and could be no basis in prior experience. This is the kind of learning that has to take place when you grab the knob of an unmarked door and try to open it. With no hints available, the door might require either a pull or a push: nothing in nature says it *has* to be either way. So what do you do to figure out how this door opens? You don't figure it out. You pull. Or you push. Whichever comes to mind first. If the door doesn't open, you have the information you need: do the opposite. If it opens, you also have the required information: don't change the way you did it. But before you could get that information, to select the right move out of the possible moves, you had to try something for *no* reason.

This is what I assume to be the basic principle of reorganization, which I could not put any better than Campbell did. Act at random, and select future actions on the basis of the consequences.

Another way of putting this is a little more systematic, and suggests at least some sort of organized system at work. Suppose we have a reorganizing system that is capable of acting on another system (of which it actually could be a part) to change the organization of that system. In this case I don't mean the organization of the *behavior* of the target system, but the very structure of that system, the physical connections in it. Changing the structure will, of course, change the behavior, but the reorganizing system doesn't act on the behavior directly. It acts on the *behaving system*. That's how Ashby's ultrastable homeostat works. It doesn't inject signals into the homeostat: it switches connections.

The reorganizing system must not only be able to alter physical connections in the target system, but it must know when to *stop* altering those connections. This is the "selective retention" part. Each change in the structure of the behaving system will alter the way that system interacts with its environment. The change in interaction will have many consequences, most of which, probably, are irrelevant to the system as a whole. Some of these changes, however, will have indirect effects on the welfare of the system itself, including the reorganizing system. These indirect effects are the basis for selection, and therefore the basis for starting and stopping the process of reorganization.

Selection necessarily implies a selection criterion. Some indirect effects of behavior are "good" and some are "bad," or at least "not good enough." But this reorganizing system has to be *dumb*. It has to work even when the system it is working on has only the barest suggestion of competence in it. It has to work without any theory, without any knowledge of an external world, without any memories of prior experience (from this lifetime, anyway).

So this system has to be told, somehow, what is good or not good enough, and perhaps even too good. It has to be given reference signals from somewhere. For lack of a better idea, I'll say DNA.

Furthermore, these reference signals have to have highly specific meanings. It won't do to posit a reference signal that says "survive!" How could a dumb reorganizing system with no linguistic capabilities know what "survive" means? It won't do to say "organize behavior." There isn't any behavior to organize at first. No, these basic reference signals have to be expressed in much more concrete terms that have direct meaning to the reorganizing system. They have to say things like "this much blood sugar" or "this body temperature" or "this carbon dioxide concentration in the blood." Of course they might also say more interesting things, like "no more than this amount of total error signal in the brain," or even "this pretty pattern of forms in your vision." We mustn't underestimate the power of a billion years of evolution. The selection criteria that make reorganization work as it does might prove to be extremely sophisticated.

But we know one thing they will not be: intelligent. Intelligence is something that gradually forms as the brain becomes organized into a hierarchy of perception and control. Intelligence is the *product*, not the cause, of reorganization. The intellectual skills found in the fully-formed adult control hierarchy are not available before it has been built. The reorganizing system has to work from the very beginning of life, so it can't take advantage of what it has not yet brought into being.

The reference signals—let's call them "intrinsic" reference signals to distinguish them from the kind in the acquired hierarchy of control—can have no effect by themselves; they are only specifications. The reorganizing system has to be able to *sense* the states of the variables that relate to the reference signals. And the sensed states have to be compared with the reference signals; the reorganizing system has to contain comparators, one for each intrinsic variable. Ashby called these intrinsic variables "critical variables." He saw the reference states as upper and lower limits, while I see them as single target values, but that's the sort of difference we might hope to settle through experiments, and isn't important here.

So we arrive at the idea that the reorganizing system is really a control system. It is, however, a very peculiar sort of control system, in that its output actions are random. It does not act "against" error: it continues to act until the error disappears. The error, of course, is simply the total absolute difference between the sensed intrinsic state and the set of all corresponding intrinsic reference signals. Ashby didn't spell out the perceptual functions or the reference signals in his ultrastable homeostat, but he did build them into it, perhaps without even realizing exactly what he had designed.

One helpful notion is the idea of *rate of reorganization*, which would be measured simply as so many changes per second, or hour. If there is a lot of intrinsic error, the reorganization rate will be high. As intrinsic error falls, assuming it does, the rate of reorganization will slow, until finally when the intrinsic error is completely gone, the reorganization will go at a rate of zero: it will stop. Another way to say it is this: if the result of a reorganization is a bigger intrinsic error, reorganize again right away. If it is smaller, delay the next reorganization.

This is exactly how the bacterium *E. coli* makes its way up chemical gradients (or down gradients of repellents). *E. coli* can't steer. It can swim in a straight line, or it can reverse its flagellae into a disorganized mess and tumble for a moment. It senses the time rate of change of concentration. If that rate of change is positive (for an attractant), the next tumble is delayed. If the rate of change is less, or negative, the next tumble occurs sooner. If *E. coli* could simply steer, it could make its way up the gradient only about 30 percent faster. And it can control in this way its relationship to 27 *different substances*.

Now, that is interesting. This miniscule creature that can locomote only by varying the interval between random tumbles quite effectively controls the sensed concentration of substances that it experiences. Furthermore, as Richard Marken and I have found through many simulations, this means of locomotion is extremely effective: it works equally well in one, two, or three dimensions, in planar gradients or radial gradients, in inverse-square gradients or linearly-declining gradients, and even in gradients arising from multiple sources (*E. coli* doesn't care where it gets its lunch). It can track moving sources, and it can overcome disturbances that push it away from the source.

It is essentially impossible to thwart this control system, except through direct superior physical force, precisely because it makes no assumptions at all about the spatial properties of the medium in which it works. The random mode of action is totally atheoretical, and could therefore work in any environment that presented a minimum degree of continuity of response to actions. On first encounter, we might think of *E. coli*'s way of navigating as very primitive and crude. On further acquaintance, however, we come to see it as the most powerful and general means of control that can possibly exist.

This, it seems to me, is exactly the kind of control system we need for a reorganizing system. Totally dumb, but extremely powerful in its ability to find its way toward zero error in any sort of environment.

Of course it may prove that even this powerful kind of primitive control system isn't enough to account for the rate at which the human hierarchy becomes organized during one life. Even more powerful principles may have to be discovered. But I believe that they will all share one property: they will work to create organization by attempting to control what matters directly to them, the organization of behavior that results being only a side-effect. This principle, I believe, has very broad applications, not only to the growth of a single organism from one cell, but of a species from the same simple origin, or from even a simpler one. But I said I was going to skip that subject, and I will.

The concept of a reorganizing system fills in a missing part of the control-system model: the explanation of how it got that way. No doubt we will find that the organism inherits many features that make this process more efficient—a predisposition to organize eleven types of control systems, although not any particular form of any type, for example. While there is much unexplained by this model—just where and how the process takes place, for one example—I think it shows us at least how an organism could become organized, maintain its organization, and acquire new organizations that pertain specifically to the continued existence of the organism in a wide variety of changing environments. I believe that I am describing the basic mechanism of the phenomenon that cyberneticists call autopoiesis.

Conclusions

What happens when more than one control system is present in a given environment—when control systems interact? One ever-present possibility is conflict. Conflict is particularly devastating when two control systems are involved, especially if both of them have very high error sensitivity. Conflict arises when the systems attempt to use the same external variable to achieve goals that imply different states of that same variable. The better the control systems are, the more effort they will generate to correct a given amount of error. So the best control systems, when they come into conflict, will immediately generate very large efforts, treating the efforts of the other system as disturbances to be counteracted. This is the genesis of human violence, because human beings are the best control systems there are.

Violence, aggression, hostility, war, murder—these phenomena do not arise from specialized inborn traits or learned habits, but simply from the normal operation of living systems that are unaware of how they, or more particularly the other systems, work. To move a rock you exert a lifting force that is at least equal to its weight: the rock moves, because it is not a control system and doesn't care where it is. To move a human being in a similarly arbitrary fashion, you might have to kill him first. And we often do that, figuratively and literally. The easiest way to deal with opposition is to overwhelm it with superior force or destroy it.

Control theory is above all a theory of living autonomous systems. Living systems are all control systems, the only natural ones, and the essence of their lives is to control what happens to them, rather than leaving their fates to wind, tide, erosion, and entropy. But the human control systems that concern us most are also very new control systems, largely ignorant of their own nature and prone to treat other living systems, including human ones, as little more than objects to be moved or disturbances to be overcome. Indeed there have been times in human history when many people saw inanimate nature as full of purposive control systems, and human beings as only passive victims of nature's intentions.

It is not easy for control systems, human beings, to live together. Even when they attempt to cooperate, they end up pitted against each other over minor differences in perception or goal. They just can't help trying to keep their own errors corrected. To be with others one has to learn deliberately to loosen the control, to lay back, to tolerate error, to be a little less skillful. To expect less, perhaps, of group efforts than of individual ones, but to value them, perhaps, more. To let reorganization ease the strain. To realize how isolated we are; how miraculous it is that we have any contact at all, mind to mind. To appreciate the vast sea of mystery that fills the space between us, through which we would have great trouble steering without the touch of other human hands on the helm, the surprise of other human thoughts about the course.

Published Works by William T. Powers on Living Control Systems, 1957-1988

1957a   (with R.L. McFarland and R.K. Clark) "A general feedback theory of human behavior," *University of Chicago Counseling Center Discussion Papers* 3(18), 21 pp.

1957b   (with R.L. McFarland and R.K. Clark) "A general feedback theory of human behavior: A prospectus," *American Psychologist* 12(7), July, 462.

1960a   (with R.K. Clark and R.L. McFarland) "A general feedback theory of human behavior: Part I," *Perceptual and Motor Skills* 11(1), August, 71-88. Reprinted in Ludwig von Bertalanffy and Anatol Rapoport, eds., *General Systems: Yearbook of the Society for General Systems Research 5*, Society for General Systems Research, Ann Arbor, 1960, 63-73. Reprinted in altered form, as "A general feedback theory of human behavior," in Alfred G. Smith, ed., *Communication and Culture: Readings in the Codes of Human Interaction*, Holt, Rinehart and Winston, New York, et al., 1966, 333-343.

1960b   (with R.K. Clark and R.L. McFarland) "A general feedback theory of human behavior: Part II," *Perceptual and Motor Skills* 11(3), December, 309-323. Reprinted in Ludwig von Bertalanffy and Anatol Rapoport, eds., *General Systems: Yearbook of the Society for General Sys-*

Prepared by Gregory Williams, with help from Mary A. Powers, Richard J. Robertson, Philip J. Runkel, and Stuart A. Umpleby. Works with italicized dates are reprinted in this volume.

tems Research 5, Society for General Systems Research, Ann Arbor, 1960, 75-83.

1971 "A feedback model for behavior: Application to a rat experiment," *Behavioral Science* 16(6), November, 558-563.

1973a *Behavior: The Control of Perception*, Aldine Publishing Co., Chicago, xiv + 296 pp.

1973b "Feedback: Beyond behaviorism," *Science* 179(4071), January 26, 351-356.

1973c (with William M. Baum and Hayne W. Reese) "Behaviorism and feedback control," *Science* 181(4105), September 21, 1114, 1116, 1118-1120.

1974a "Some cybernetics and some psychology," *ASC Forum* 6(4), Winter, 4-9.

1974b (with Paul Bohannan and Mark Schoepfle) "Systems conflict in the learning alliance," in Lindley J. Stiles, ed., *Theories for Teaching*, Dodd, Mead & Co., New York, 76-96.

1974c "Applied epistemology," in Charles D. Smock and Ernst von Glasersfeld, eds., *Epistemology and Education: The Implications of Radical Constructivism for Knowledge Acquisition*, Mathemagenic Activities Program Follow Through Research Report (14), University of Georgia, Athens, 84-98.

1975 Review of *The Logic of Social Systems* by Alfred Kuhn, *Contemporary Sociology* 4(2), March, 92-94.

1976a "Control-system theory and performance objectives," *Journal of Psycholinguistic Research* 5(3), July, 285-297.

1976b "The cybernetic revolution in psychology," *ASC Cybernetics Forum* 8(3 & 4), Fall/Winter, 72-86.

1976c "Reply to Katz' analysis," *ASC Cybernetics Forum 8* (3 & 4), Fall/Winter, 143-146.

1977 "Rückkopplungsprinzipien in der Organisation von Verhalten," in H. Zeier, ed., *Pawlow und die Folgen: Von der klassischen Konditionierung bis zur Verhaltenstherapie, Die Psychologie des 20. Jahrhunderts 4*, Kindler Verlag AG, Zürich, 573-613.

1978 "Quantitative analysis of purposive systems: Some spadework at the foundations of scientific psychology," *Psychological Review 85*(5), September, 417-435.

1979a "A cybernetic model for research in human development," in Mark N. Ozer, ed., *A Cybernetic Approach to the Assessment of Children: Toward a More Humane Use of Human Beings*, Westview Press, Boulder, 11-66.

1979b "Cause/effect metaphors versus control theory," *The Behavioral and Brain Sciences 2*(1), March, 115.

1979c "Cybernetics and the assessment of children," *ASC Cybernetics Forum 9*(1 & 2), Spring/Summer, 57-65.

1979d "Degrees of freedom in social interactions," in Klaus Krippendorff, ed., *Communication and Control in Society*, Gordon and Breach Science Publishers, New York, London, and Paris, 267-278.

1979e "The nature of robots: Part 1: Defining behavior," *BYTE 4*(6), June, 132, 134, 136, 138, 140-141, 144.

1979f "The nature of robots: Part 2: Simulated control system," *BYTE 4*(7), July, 134-136, 138, 140, 142, 144, 146, 148-150, 152.

1979g "The nature of robots: Part 3: A closer look at human behavior," *BYTE 4*(8), August, 94-96, 98, 100, 102-104, 106-108, 110-112, 114, 116.

1979h "The nature of robots: Part 4: Looking for controlled variables," *BYTE* 4(9), September, 96, 98-102, 104, 106-110, 112.

1980a "A systems approach to consciousness," in Julian M. Davidson and Richard J. Davidson, eds., *The Psychology of Consciousness*, Plenum Press, New York and London, 217-242.

1980b "Pylyshyn and perception," *The Behavioral and Brain Sciences* 3(1), March, 148-149.

1981a "Foreword," in William Glasser, *Stations of the Mind: New Directions for Reality Therapy*, Harper & Row, Publishers, New York, et al., ix-xiii.

1981b "Maximization, or control?" *The Behavioral and Brain Sciences* 4(9), September, 400-401.

1984 "Interactionism and control theory," in Joseph R. Royce and Leendert P. Mos, eds., *Annals of Theoretical Psychology 2*, Plenum Press, New York and London, 355-358.

1985 (with Richard J. Robertson) *Introduction to Modern Psychology: General Psychology: The Control Theory View*, Richard J. Robertson, Chicago, ii + 195 pp.

1986a "Greetings from Bill P.," *Feedback: The Control System Group Newsletter*, March 5, 1-2.

1986b "Intentionality: No mystery," *The Behavioral and Brain Sciences* 9(1), March, 152-153.

1986c Summary of presentation at August 1986 CSG meeting, *Control System Group Newsletter*, October 15, 1-2.

1986d "Interaction: The control of perception: Commentary on a paper by J.M. White," in *Center for Systems Research Working Paper* (86-4), University of Alberta, Edmonton, November, 1-9.

1986e "On purpose," *Continuing the Conversation: A Newsletter on the Ideas of Gregory Bateson* (7), Winter, 1-3.

1986f (with Ernst von Glasersfeld) "A control conversation," *Continuing the Conversation: A Newsletter on the Ideas of Gregory Bateson* (7), Winter, 3-4.

1986g (with Larry Richards) "Letters," *Continuing the Conversation: A Newsletter on the Ideas of Gregory Bateson* (7), Winter, 4-6.

1987a "Foreword," in Edward E. Ford, *Love Guaranteed: A Better Marriage in Eight Weeks*, Harper & Row, Publishers, San Francisco, et al., ix-xiv.

1987b "Is control theory just another point of view?" *Continuing the Conversation: A Newsletter of Ideas in Cybernetics* (9), Summer, 1-2.

1987c "Thoughts for this year's meeting," *Control System Group Newsletter*, July 24, 2.

1987d "Control theory and cybernetics," *Continuing the Conversation: A Newsletter of Ideas in Cybernetics* (11), Winter, 13-14.

1988a "The asymmetry of control," *Control System Group Newsletter*, February, 3.

1988b "A nonfunctional analysis of behavior," *The Behavioral and Brain Sciences* 11(1), March, 143-144.

1988c "Thank you, friends," *Control System Group Newsletter*, March 3, 1.

1988d "An outline of control theory," in *Conference Workbook for "Texts in Cybernetic Theory": An In-Depth Exploration of the Thought of Humberto R. Maturana, Wil-*

liam T. Powers, and Ernst von Glasersfeld, American Society for Cybernetics, Felton, California, October 18-23, 1-32.

1988e "*Comment:* The trouble with S-Delta," *Continuing the Conversation: A Newsletter of Ideas in Cybernetics* (14), Fall, 5-6.

1988f "From Bill Powers:" *Control System Group Newsletter*, November, 2.